SCIENCE SNIPPETS

Brief Columns From a Small Town Newspaper

MAX SHERMAN

outskirts
press

Science Snippets
Brief columns from a small town newspaper
All Rights Reserved.
Copyright © 2020 Max Sherman
v3.0

The opinions expressed in this manuscript are solely the opinions of the author and do not represent the opinions or thoughts of the publisher. The author has represented and warranted full ownership and/or legal right to publish all the materials in this book.

This book may not be reproduced, transmitted, or stored in whole or in part by any means, including graphic, electronic, or mechanical without the express written consent of the publisher except in the case of brief quotations embodied in critical articles and reviews.

Outskirts Press, Inc.
http://www.outskirtspress.com

ISBN: 978-1-9772-2512-2

Cover Photo © 2020 www.gettyimages.com. All rights reserved - used with permission.

Outskirts Press and the "OP" logo are trademarks belonging to Outskirts Press, Inc.

PRINTED IN THE UNITED STATES OF AMERICA

TABLE OF CONTENTS

3 D printing 1
　—What will the future hold!
A Backpacker from the Ice Age 4
A trip to the dentist and biofilms 6
Aging 9
　—A relentless process.
Algae 13
　—A life saver!
Alzheimer's 17
　—The most dreaded age-related disease.
Anesthesia 21
　—Questions Still Abound!
Annals of Improbable Research 25
Ants 28
　—The real masters of our planet.
Bacteriophage therapy 32
　—Getting closer?
Bed bugs 36
　—Creatures of the night!
Bees 40
　—Badly needed but vanishing
Bill Bryson 44
　—A writer worth reading
Copying nature's ways 48
98.6 F 52
　—It is not the average body temperature anymore!
The breath of life 56
Cancer 59
　—Prevention may be the only solution!

Candida auris 63
   —A mysterious and fatal yeast
Carbon 67
   —We wouldn't be here without it.
Charles Krauthammer 71
   —Physician, historian, commentator—a great loss!
Charles Sherrington (1857-1952) one of a kind. 75
Cicero 79
   —An ancient Roman worth emulating!
Circadian Rhythms 83
   —Another of nature's wonders.
Curiosity 87
   —The most valuable commodity!
David Nachmansohn, nerve gases, and electric eels. . 91
Direct to Consumer (DTC) Drug Ads 95
Doctor Fish 99
   —A cure for psoriasis?
Drugs for elderly patients 103
   —The Beers Criteria
Dying 107
   —Ssomething to think and not worry about.
Elephants 111
   —Large, leathery, lumbering, lovable animals.
Emery Andrew Rovenstine MD 115
   —Someone to be proud of.
Escherichia coli 119
   —Friend or foe?
Falls in the Elderly 123
Clostridium difficile colitis 127
   —A very unusual treatment method.
More smart men 131
Frances Oldham Kelsey and Helen Brooke Taussig . 135
   —Two memorable advocates for drug safety

Francis Galton 139
    —Another polymath!
Fruit Flies 143
    —Pioneers in genetic research!
The Giraffe 147
    —Magnificent but endangered!
Hand washing 151
    —A New Year's Resolution worth keeping!
Happiness 156
    —Something we all desire.
Harvey Washington Wiley and the Pure Food and Drug Act 159
HM and HeLa 163
    —Letters you may wish to learn more about!
Hibernation 167
    —A bear's secret to good health.
Our microbiota 171
    —An essential companion!
Human Errors 175
Poliomyelitis (Polio) 178
    —Still hanging around.
Human challenges 182
    —How long has this been going on?
Isaac Newton 186
    —One of a kind!
Itching 190
    —A diabolical, peculiar and often unsolvable problem!
Knots 194
    —A tied and true hobby!
Leeches 198
    —Still used in medicine

Leprosy 201
   —*Not yet a thing of the past.*
Lightning Bugs 205
   —*Vanishing memories of youth!*
Locusts 209
   —*A humanitarian crisis!*
Logic and rhetoric 213
   —*Two essential courses well worth studying.*
Mathematics and Numbers 216
   —*Shouldn't we be more numerate?*
Measles 220
   —*Back again!*
Medical Quackery 224
   —*Not just something from the past.*
Medical Writers 228
The Mediterranean Diet 232
   —*Worth adopting.*
Menstruation 236
   —*Still a mystery?*
Adopting Mindfulness Training 240
   —*Why not?*
Mosquitoes 244
Natalie Angier 248
   —*A writer worth reading.*
The day the dinosaurs died 252
New Discoveries in Human Anatomy 255
Thanksgiving 259
   —*What to be thankful for.*
The surprising octopus 263
   —*One of nature's marvels!*
Otto Warburg 267
   —*Another medical genius.*
Examining Oetzi the Iceman 271

Oxygen	275
—*The subject of this year's Nobel Prize*	
Oxytocin	279
—*A surprisingly versatile hormone.*	
The Pancreas	283
—*Mysteries of a Hidden Organ*	
Passenger pigeons	287
—*What happened?*	
Patient counseling	291
—*Ask your pharmacist!*	
Philosophy	295
—*A subject to know more about*	
Poems	299
—*Why do I find them so inscrutable?*	
Another polymath	303
—*John Von Neumann*	
Polypill	307
—*A means to live longer?*	
Potatoes	311
—*Are they becoming an endangered species?*	
Probiotics and Microflora	315
Pythons	325
—*A model to study digestion and human heart disease.*	
Rats	328
—*Built to last!*	
Roundworms	332
—*Successors to the fruit fly?*	
Salt	335
—*Widely used but often misunderstood.*	
Scorpions	339
—*Scary and nasty creatures!*	

Sleep 343
    —*Still a mystery!*
The sense of smell 347
    —*Dogs or us?*
How important it is to hear! 351
Spanish Flu 355
    —*May it never return.*
Spiders 359
    —*Vile but valuable!*
Stanley Cohen 363
    —*He made a breakthrough in cellular growth!*
Stanley Falkow 367
    —*Microbiologist extraordinary.*
Strange unexplained diseases 371
Stuart Levy 375
    —*A pioneer in the proper use of antibiotics*
Super Navigators 379
    —*Getting around without GPS!*
Termites 382
    —*Surprising facts!*
The brain 386
    —*Much still to be learned and unlearned.*
The Doomsday Clock 389
The Epizootic of 1872 393
    —*A surprising series of events!*
♥ The Heart 397
    —*A symbol and its function.*
The Periodic Table 400
    —*Something to know more about.*
The Science of Cooking 404
    —*Make it more fun.*
The Scientific Revolution 408

The two smartest men who have lived in the past 400 years. 412
Ticks 416
— *Much more dangerous than we thought.*
Touch 420
— *A new science?*
Tuberculosis 424
— *A continuing global problem*
Vanishing birds 428
— *Something to be alarmed about!*
Reptiles, bugs and marine life used for drugs 432
Viruses 436
— *Miraculous life forms.*
Vision 439
— *What a wonderful gift!*
Visualizing data 443
Watching worthwhile television 447
Water 451
What is science? 455
Zebrafish 458
— *An amazing tool for biomedical research*

3 D printing

—*What will the future hold!*

3 D printing or additive manufacturing is a way of making objects in which a special printer with a digital blueprint, can produce an item in three dimensions. In contrast, and for centuries, the world's manufacturing has been deemed either subtractive or formative. Subtractive is best suited for parts with relatively simple geometries and are created by cutting, drilling, chiseling, chopping, scraping or carving. These methods are used with functional materials (metals or plastic) and while useful, create loads of waste, scrap and leftovers and extract even more valuable raw materials from the earth. Formative manufacturing is used for high volume production of the same parts and require a large initial investment in tooling or molds. Additive manufacturing methods are relatively new, they are best suited for low volume, complex designs where formative or subtractive methods are unable to product unique one off rapid prototypes. More uses are in the offing, however.

3 D methods of manufacturing are now becoming widely used in the medical device and pharmaceutical industry and the FDA has issued guidance for manufacturers to ensure that parts are tested for function and durability

and made in accordance with good manufacturing practices. There are now more than 100 such devices currently on the market. These include patient matched devices tailored to fit a patient's anatomy, examples include knee replacements and implants designed to fit like a missing puzzle piece into a patient's skull for facial reconstruction. The first drug produced on a 3 D printer is now used to treat seizures and has a more porous matrix than the drug manufactured in the traditional way, enabling the drug to dissolve more rapidly and to work faster. Burn patients in the near future will be treated with their own new skin cells that are 3 D printed onto their burn wounds. There is a potential for 3 D printing to be eventually used to development replacement organs.

Diane Ackerman, in her excellent book, the Human Age, describes 3 D printing as "following a blueprint as if it were a musical score, a nozzle glides back and forth over a platform, depositing one microscopic drop after another in a molten fugue, layer upon layer until the desired object rises like a sphinx from the sands of disbelief. Aluminum, nylon, plastic, chocolate, carbon nanotubes, soot, polyester—the raw material doesn't matter, provided it is fluid, powder or paste." Again, according to Ms. Ackerman, like many other technologies, 3 D printing has a potential dark side, people have printed out hand guns, brass knuckles, and skeleton keys than can open most police handcuffs. Fortunately, future laws will undoubtedly restrict access to illegal and patented blueprints, and also to dangerous metals and gases, explosives, weapons, and street drugs. On the positive side, 3 D printing could fabricate artifacts that cannot be created in any other way, and who knows what artworks and breakthroughs will emerge.

Today, because 3 D printing is somewhat a novelty, we value its products highly, but when cheap 3 D printers become more common, and factory 3 D printing replaces the assembly lines and warehouses, we'll live in an even more improbable world. As cars, rockets, furniture, food, medicine, musical instruments and much more become readily printable, it is bound to temporarily or even permanently unnerve the world's economy. Is it possible that companies won't need to hire scores of workers, buy raw materials, ship or stock or produce anything? Ms. Ackerman believes 3 D printing companies will be extremely profitable, because their overhead will be much lower, and they will sell only the clever designs or raw materials. In all likelihood, 3 D printing will eventually revolutionize life from manufacturing to art, and usher in the next great cultural and psychological age. Forbes magazine believes 3 D printing will just like the industrial revolution, the assembly line, the advent of the internet, and the social media phenomenon and become a true game changer. Evidence already exists to prove Forbes correct, 3 D printers are being used to fabricate hard to find parts for classic cars, the Smithsonian uses a 3 D printer to build dinosaur bones, and NASA uses it to build a prototype of a two man Space Exploration vehicle. Boeing is 3 D printing seven hundred parts for its fleet of 747's, and it has already installed twenty thousand such parts on military aircraft. 3 D printed cars are next. With 3 D printers, complexity is free, and for the first time making something complicated with fine detail or ornate features is not harder than making a spoon or a paper weight. In a sense, history is being made in our contemporary world.

A Backpacker from the Ice Age

~~~

Jubal…was the father of all who play harp and flute…
Tubal-Cain…forged all kinds of tools out of bronze
and iron.

<div align="right">Genesis 4:21-22</div>

HIKERS IN 1991 came upon a strange sight in the Italian-Austrian Alps. Thawing from a boundary of glacial ice were the mummified remains of a man from the Old Testament era. Given the name Ötzi and called "The Iceman," this rugged outdoorsman lived during the ice age which occurred in the centuries following the global Genesis flood. This colder period dates around 2000 B.C., an era which includes the contemporaries Abraham, Job and also Noah who himself lived 350 years past the flood. Ötzi is dated by some aa a thousand years earlier, living around 3000 B.C.

The physical details of the Iceman are remarkably modern. Ötzi's waterproof shoes are made of layers of bearskin and deer-hide with leather shoelaces. His coat and leggings are stitched from several animal hide fragments. There is also a bearskin cap with a leather chin strap. The man's mummified skin shows dozens of tattoos, including a cross symbol which is of special interest in this early pre-Christian era.

Several items of survival gear also were found nearby Ötzi. These include a flint knife, copper axe, and a bow and quiver with a dozen feathered arrows. A backpack holds medicinal herbs and also a fire-making kit with flint, pyrite and tinder.

What is the point of this description? It is simply that Ötzi defies the common assumption that early people were primitive and animal-like. Ötzi closely resembles a modern hiker who was well equipped for the cold environment. Ever since Adam and Eve, mankind has been blessed by God with abilities and intellect. Our Bible verse describes the music and metallurgy abilities of some of the earliest people on earth. As we hike today we carry on the grand tradition of enjoying nature up close. Often, the modern trails we hike were first traveled by native American Indians in centuries gone by.

Detailed analysis shows that Ötzi probably died from blood loss due to an arrow point found lodged in his shoulder. His demise appears to be a murder mystery from the past, a sad commentary on fallen human nature.

A TRIP TO THE DENTIST AND BIOFILMS.

I TRY TO visit the dentist three or four times a year and each time I go the hygienist begins to chip away at the inordinate amount of tartar that adheres to bridges, caps, and all of my remaining natural tooth surfaces. It puzzles me, for despite water picking, diligent brushing and flossing, the stuff builds up unmercifully. I finally decided to learn more about my dilemma and in my research rediscovered that plaque is the precursor to tartar and is defined as the soft, sticky film that builds up on teeth and contains millions of bacteria. Plaque, if not removed daily, can eventually harden into tartar which without flossing and brushing is most difficult to remove. Left alone, tartar can eat away tooth enamel by producing lactic acid, and eventually result in periodontal (gum) disease and tooth loss.

Further study on the "tartar mystery" indicated that plaque and tartar are considered to be "biofilms." A biofilm can be defined as a community of microorganisms (bacteria) that are associated with a surface and typically enveloped in a complex architecture. The ability to form biofilms is one of the most remarkable characteristics of bacteria. Steps needed for formation are as follows: the free swimming bacteria (the mouth contains more than 700 species) alight of a surface, arrange in clusters and attach; the cells begin producing a slime like extracelluar matrix; the cells signal one another to form a colony; chemicals are produced which literally promote the existence of a new species with genetic changes and an altered metabolic state. Some cells may return to their free living state, escape and form new colonies. In the natural environment and in most clinical and industrial settings, biofilms are composed of many different species, sometimes numbering in the hundreds. Sometimes one species feeds on the metabolic wastes of another, aiding them both. All this is possible because of the unexpected high degree of a coordinated cell-to-cell interaction. There must be some signaling that monitors the density of the population of bacteria. This complicated process is known as "quorum sensing." Bacteria are a lot smarter than we give them credit for.

In addition to tartar, there are many other examples of biofilms. In fact, it has been estimated that up to 65 percent of all human microbial infections involve biofilms. Bacteria that form biofilms have an increased resistance to antibiotics and protection from the body's ability to fight off an infection. This is of particular importance in light of the explosive growth in the use of indwelling medical devices including catheters, stents, shunts, heart valves, dental implants and total joint prostheses. Materials selected to fabricate these devices have mechanical properties that allow them to perform their function without causing trauma to surrounding tissues. They are not inert, however, and their surfaces are capable of physical, chemical and biological interactions. In addition, unlike our immune system, they have none of the defense mechanisms that protect the surfaces from bacterial colonization. Surface irregularities can also induce entrapment of bacterial cells.

Catheters are particularly prone to the effects of biofilms. One researcher who examined 57 catheters that had been in place for one to fourteen days found that bacteria could be recovered from the surface of 54 percent of the devices and that in over half of the cases, two or more organisms were discovered. The most alarming problem is that persistent infections associated with biofilms are often resistant to antibiotics. In fact, when encased in biofilms in the human body, bacteria are a thousand times less susceptible. One example is cystic fibrosis, a disease characterized by hordes of bacteria that form a sticky film on the surface of a patient's lungs. Once these bacteria attach, drugs won't kill them. Moreover, the protective mechanisms behind biofilm resistance are still mostly unknown. Biologists do, however, now understand much more about the formation of bacterial biofilms and therefore it may be possible in the future to find drugs that target their unique properties and inhibit their formation.

In summary and in plain English—- biofilms are nasty once they form, and incredibly difficult to get rid of. Fortunately, even though plaque constantly grows and tartar continues to form both can be removed by brushing, water picking and flossing every day. Visiting your dentist at least once a year for professional cleaning and oral exams is important too. There is at least one more benefit—there is recent evidence that preventing gum disease diminishes the risk of cancer in older women and likely a similar effect in older men.

Max Sherman, Warsaw, IN.

Aging

—A relentless process.

I HAVE TAKEN a keen interest in the process of aging and all that it entails the older I become. One reason may be the change in attitude regarding the perception of what it means to all of us. William Osler (1849-1919), once hailed as the most illustrious physician in our history, remarked that men older than 60 should be retired and suggested that men passed that age should be chloroformed or in other words, terminated. Because I am much older, I obviously and vehemently disagree with the doctor's philosophy and one of the reasons I collect books on the subject and continue to search for ways to slow down the aging process. Today, aging is a timely and fashionable topic as America has become an increasingly "gray" nation with people over 85 comprising the fastest growing segment of the population. Aging, of course, brings on a host of infirmities and despite all we do to care for ourselves, many of us lucky (?) enough to get older may spend our last days residing a nursing home.

Before I disclose a few tips from the literature on anti-aging, it is important to review process itself. Usual aging involves two sets of problems, First, many body organs, including the kidneys, heart, and lungs, gradually lose strength with advancing age, immune function also declines. These changes place the elderly at risk of disease and dysfunction, especially when major stress occurs. The second set of problems with usual aging relates to the buildup of blood fats and sugar and subsequent hypertension. All of the major organ systems are affected. There is little doubt the incidence of heart disease increases exponentially with age. It is the leading cause of death in older age groups in industrialized nations. Atherosclerosis, a thickening and hardening of the walls of the arteries is responsible for most deaths. Gerontologists (specialists in aging) believe that the incidence of cancer with age is the result of an immune system less capable of detecting and destroying cancer cells. Hormonal changes with age reduce the physiological reserve in tissues and organs. The secretion of many hormones including testosterone, insulin, and thyroid, decrease with normal aging. Moreover, there is general agreement the weight of the brain decreases from young adulthood to old age. The number of cells may not of itself be significant, however if the loss occurs in cells whose vital functions cannot be assumed by other cells, the loss could be. Alzheimer's disease, a condition characterized by a mass of fibers found in brain cells is a disease of older people. There is also loss of bone with aging, it begins at around fifty for both sexes, but then proceeds more rapidly in women.

There are many theories as to why we age. Some focus on the accumulation of errors in the genetic code, others invoke the loss of telomeres which are the repeated sequences found at the end of chromosomes. Telomeres shorten with repeated cell divisions and cells that are normally replaced no longer do so. Another theory is that over time reactions between proteins and sugars in the body form compounds which can damage other proteins as well as DNA. Cross linking occurs as does mutations and there is reduced energy which disrupts vital organs. Free radicals are the cause.

There are a number of manufacturers of nutritional products that promote antioxidants as a means to retard aging. These compounds including vitamins E and C and butylated hydroxytoluene are alleged to work by combating the effects of free radicals. None have been proven to be safe and effective. Another product, resveratrol, which is found naturally in the skin of red grapes, is another promising candidate. There have been a half dozen or do other medications or supplements that appear to improve damage within our cells and thus help prolong life. This includes an anticancer drug and a popular diabetes drug, metformin, which seems to lower the incidence of cancer and provides beneficial effects for the heart. Rapamycin, a drug used to coat coronary stents, has shown beneficial effects on mice which appear to stay healthier and more youthful when administered the drug. Another method to retard aging stems from dietary restrictions. Experiments as far back as the 1930s have shown that feeding laboratory animals about a third fewer calories can dramatically increase longevity.

While the cure for aging may be a long way off or may never occur, we do know that aging is affected by lifestyle. Men and women aged 45 to 79 who are physically active, eat plenty of fruits and vegetables, do not smoke and consume alcohol moderately have an average one-fourth the risk of death during any given year than people with unhealthy habits. The drop in mortality risk among those with healthier habits was equivalent to being 14 years younger. It is like turning the clock back 14 years—-a worthy goal for all of us, although Dr. Osler might disagree. Finally, in a recent book on successful aging, the author suggests that "we need to keep moving, keep happy, keep learning and keep connected."

Algae

—*A life saver!*

If you own a swimming pool or are acquainted with the sea you no doubt are familiar with algae. These simple plants are diverse and found almost anywhere on the planet, ranging from microscopic to large seaweeds. Some giant kelp can grow to more than 100 feet in length. Microalgae include both cyanobacteria (similar to bacteria, and formerly called blue-green algae) as well as green, brown and red algae. Most algae grow through photosynthesis—by converting sunlight, carbon dioxide and a few nutrients, including nitrogen and phosphorous, into material known as biomass. Algae play an important role in many ecosystems, including providing the foundation for the aquatic food chains supporting all fisheries in the oceans and inland, as well as producing about 70 percent of all the air we breath. They work their combinatorial magic, by burping oxygen.

Take a breath: most of the oxygen we inhale is made by algae. What is waste to them is priceless to all respiring animals. Without algae, we would gasp for air. There is no shortage of algae. The oceans are blanketed in a dense but invisible six-hundred-foot-thick layer of them. There are more algae in the oceans than there are stars in all the galaxies in the universe. Swallow a single drop of seawater, and you could easily down several thousand of these unseen beings. According to Ruth Kassinger, in her new book *Slime,* they are the essential food of the microscopic grazing animals at the bottom of the marine food chain. If all algae died tomorrow, then all familiar aquatic life—from tiny krill to whales—would quickly starve. In fact, if algae hadn't evolved more than 3 billion years ago and oxygenated the atmosphere, multicellular creatures would never have graced the oceans. It was a species of green algae that, 500 million years ago, acclimated to life on land and evolved into all of Earth's plants. Without plants to eat, the first marine animals that wriggled out of the water 360 million years ago would never have survived or continued to evolve and diversify into all the land-living creatures we know today, including us. If, several million years ago, our ancestor hominins hadn't had access to fish and other algae-eating aquatic life—and thus to certain key nutrients—we would never have evolved our outsize brains.

Algae are usually thought of as plant like as there is a similarity. Each produces the same storage compounds to act as defense strategies against predators and parasites. Plants, however, show a high degree of differentiation, with roots, leaves, stems, and vascular tissues. Algae do not have any of these features, even though many seaweeds are plant-like in appearance and some of them show specialization and differentiation of their vegetative cells. However, they do not form embryos, their reproductive structures consist of cells that are all potentially fertile and lack sterile cells covering or protecting them. Moreover, algae occur in dissimilar forms such as microscopic single cells, macroscopic multicellular loose or filmy conglomerations, matted or branched colonies, or more complex leafy or blade forms, which contrast strongly with uniformity in vascular plants.

The term algae refers to macroalgae and a highly diversified group of microorganisms known as microalgae. Estimates of the number of living algae varies from 30,000 to more than 1 million species, but most of the reliable estimates refer to the numbers given in AlgaeBase. This program currently documents 32,260 species of organisms generally regarded as algae of an estimated 43,918 described species of algae. There may be more than 28,500 species waiting for description.

One type of algae that Floridians are acquainted with is *Karenia brevis*. This common marine microorganism blooms when exposed to sunlight, warm water and phosphorus or nitrates. The result is a toxic sludge known as red tide, which depletes oxygen in water, poisons shellfish and emits a foul vapor strong enough to irritate the lungs. Its frequency and severity have worsened thanks to pollution and rising water temperatures. Red tides started appearing everywhere as early as the late 18th century and early 19th centuries. The worst red tide invasion occurred in 1946-47. It started in Naples, Florida and spread all the way to Sarasota, hanging around for18 months destroying the fishing industry and making life unbearable for residents. A 35 mile long stretch was so thick with rotting fish carcasses that the government dispatched Navy warships to try to break up the mass. A similar event occurred in 2017-8, covering 145 miles of Floridian coastline.

Just recently, scientists, using satellite imaging, have discovered a record breaking belt of brown algae stretching from West Africa to the Gulf of Mexico. It may weigh more than 20 million tons and provide habitat for turtles, crabs and birds and producing oxygen for the environment. On the other hand, this seaweed makes it hard for other marine species to move and breathe. There is still much to learn about algae.

ALZHEIMER'S

—*The most dreaded age-related disease.*

It can be argued that among the number of illnesses associated with aging including osteoarthritis, osteoporosis, cancer, diabetes, cardiovascular and Parkinson's disease, none is more dreaded than Alzheimer's disease. The affected patient suffers with a progressive loss of cognitive awareness , his or her self dependence, and eventual death within 3 to 9 years after diagnosis. It is a disease where the mind slowly vanishes leaving a empty shell of a person behind. More troubling is the fact that even after years of research, there remain more questions than there are answers. This is not surprising as the brain continues to be the most complex structure in the universe. There is still much to be learned about its function. While the research continues, all of us should expect to live a long and fruitful life, but it is troubling to know that Alzheimer's disease may be the inevitable consequence.

The disease, later described in his name, was first discussed in Tubingen, Germany by Dr. Alois Alzheimer (1864-1915) in 1906 at a psychiatry meeting. He described a peculiar brain abnormality in one of his patients who died of an unusual mental illness. The autopsy revealed a thinner cerebral cortex than commonly seen in the elderly together with his findings of tangled bundles of fibers (now called neurofibrillary tangles) and abnormal clumps (now called amyloid plaques). During his presentation, Alzheimer made the assertion that the patient's dementia was likely due to these lesions. His speech was followed by a publication the following year under the title "A characteristic serious disease of the cerebral cortex." His patient's case marked the very beginning of Alzheimer's disease research. The term "Alzheimer's disease" was coined by Emil Kraepelin in 1910 and first appeared in print in his book the Handbook of Psychiatry.

As mentioned above, the principal risk factor for Alzheimer's disease is age. It is a specific disease that affects about 6 percent of the population over 65 years of age. The incidence doubles every five years after that age, with the diagnosis of 1275 new cases per year per 100,000 persons older than 65. The odds of receiving the diagnosis of Alzheimer's after 85 years of age exceed one in three. As the aging population increases, the prevalence will approach 13.2 to 16 million cases in the United States by mid-century—an alarming forecast.

Diagnosis

A diagnosis of Alzheimer's disease is most commonly made by an individual's primary care physician. The physician obtains a medical and family history, including psychiatric history and history of cognitive and behavioral changes. Ideally, a family member or other individual close to the patient is available to provide input. The physician also conducts cognitive tests and physical and neurologic examinations. In addition, the patient may undergo magnetic resonance imaging (MRI) scans to identify brain changes that have occurred so the physician can rule out other possible causes of cognitive decline. Unfortunately, understanding and effectively treating Alzheimer's disease and other dementias may be the most difficult challenge for the medical/scientific community.

Although scientists do not know how the Alzheimer's disease process begins, it appears likely that damage to the brain starts a decade or more before problems become evident. During the preclinical stage of the disease, people are symptom-free, but toxic changes are taking place. Abnormal deposits of proteins form amyloid plaques and neurofibrillary tangles throughout the brain, and once healthy neurons begin to work less efficiently. Over time, neurons lose their ability to function and communicate with each other, and eventually die. The damage begins in the cortex (a sliver of tissue behind the ears, toward the middle of the brain). It spreads to a nearby structure in the brain, the hippocampus, which is essential in transforming daily experience into lasting memories. As more neurons die, affected brain regions begin to shrink significantly. Many molecular lesions have been detected in Alzheimer's disease, but the overarching theme to emerge from the data is that an accumulation of misfolded proteins in the aging brain results in oxidative and inflammatory damage, which in turn leads to energy failure and nerve dysfunction.

Because Alzheimer's disease is complex, current treatment approaches and research focus on several aspects. They include helping patients maintain mental function, manage behavioral symptoms, and attempt to slow or delay eventual memory loss. There are four medications approved for treatment. Donepezil may be the most popular, and often combined with memantine. They may help maintain thinking, memory and speaking skills, and certain behavioral problems. The drugs, however, do not prevent the underlying disease process and may help only for a limited time. Drugs and other supportive treatment are also used to manage other common symptoms including sleeplessness, agitation, wandering, anger and depression. New methods are desperately needed to improve the process of therapy development and increase the likelihood of success.

ANESTHESIA

―∞―

—*Questions Still Abound!*

I CAN RECALL that anesthetics were the drugs that I was most intrigued by during my pharmacology classes in pharmacy school. One reason may have been their longevity—as we learned that anesthetics were used in ancient times by the Egyptians and early Greeks. Now more than fifty years after graduation, I have experienced anesthesia for the first time during cataract surgery. This direct exposure has rekindled my interest as it relates to the following questions: What is it about these drugs that bring on such a rapid response, the analgesia (pain relief) and amnesia? What is the mechanism of action— how do they work?

Throughout history, people have sought ways to relieve suffering. Many substances that control pain were found serendipitously, sometimes by trial and error. As early as 4200 B.C. people discovered plants and plant roots that could cause unconsciousness, and they were used to relieve pain. Important drugs in antiquity included hashish, mandrake and opium. All were employed into the Middle Ages and beyond, but it was not until the first half of the 19th century that anesthetics began to be used safety and effectively in surgical operations. Prior to that time surgery was traumatic, barbaric and more often fatal.

Nitrous oxide (N_2O) or laughing gas, was used in 1844, in a demonstration by an American dentist, Horace Wells, before a skeptical audience of doctors. Unfortunately the patient was unusually resistant and laughter not anesthesia ensued. Prior to Well's medical application, the main use of nitrous oxide was in traveling medicine shows and carnivals. Nitrous oxide remains useful as a general anesthetic, but in light of its low potency it is administered in combination with other drugs.

The first true anesthetic experiment occurred on October 16, 1846. Dr. John Collins Warren performed the initial surgical procedure using ether to prevent pain. The drug was administered using an inhaler apparatus designed by William Morton, who like Dr. Wells was also a dentist. He had extracted teeth under similar circumstances, without knowledge of the patient. Dr. Morton participated in the surgery and by virtue of the results had single handedly proven that ether when inhaled in the proper dose, provided safe and effective anesthesia. The operation was concluded by Warren with the words: "Gentlemen, this is no humbug." Dr. Henry Bigelow described the surgery in a report that same year. (The report became, what today's editors believe, to be the most important article published in the New England Journal of Medicine over the last two centuries.) Shortly thereafter physician and poet Oliver Wendell Holmes Sr., coined the term "anesthesia" to describe the drug-induced insensibility to sensation.

James Young Simpson (1811-1870) a Scottish obstetrician and gynecologist whose search for a better anesthetic than ether led him to introduce chloroform. According to tradition, he and his assistants had been testing a series of chemicals when late one evening somebody knocked over a bottle of chloroform. When Simpson's wife brought dinner to the laboratory that evening she found the entire staff sleeping peacefully in strange positions. The breakthrough for chloroform came in 1853 when it was used by Dr. John Snow to deliver Queen Victoria's son, Prince Leopold. Because of its high toxicity, chloroform is no longer used as an anesthetic.

Today, halogenated ethers have replaced most other compounds for use as inhalation anesthetics. Examples include isoflurane, enflurane, desflurane, and sevoflurane. Halogenated ethers have the advantage of being nonflammable and less toxic than earlier general anesthetics.

Many modern anesthetics share the same structural and clinical effects of ether. Anesthesia has evolved into a sophisticated art form. Today's general anesthetics are the most potent depressors of the brain and spinal cord used in medicine. The primary site of action for inhalation anesthetics is the central nervous system, where they inhibit nerve transmission by a mechanism distinct from that of local anesthetics. (Local anesthetics block nerve transmission to pain centers in the central nervous system.) The patient loses awareness but the vital physiologic functions such as breathing and maintenance of blood pressure, continue to function. There is general insensibility to pain. Onset of action is remarkably fast—-about 10 to 20 seconds produces an unconscious state.

General anesthetics cause a reduction in nerve transmission at synapses, the sites at which neurotransmitters are released and exert their initial action in the body.

Although a great deal of research has been done to discover physiologic effects and the sites of action, scientists still do not know the exact molecular mechanisms of action for general anesthetics or what lasting effects they may have even though these drugs are used in the most common medical procedures. One reason is that volatile anesthetics, unlike most of the drugs used in medicine, bind only weakly to these sites. Thus I must continue to wait for answers to the questions posed at the beginning of this article. I hope I am fully conscious when that day arrives.

Annals of
Improbable Research

ANYONE WITH A sense of humor and a predilection for science will enjoy a visit to the Annals of Improbable Research website and the annual list of Ig Nobel awards. The awards are for research that first makes people laugh, then makes them think and then spurs their curiosity. They are presented by Annal's science humor magazine, the Harvard Radcliffe Science Fiction Association, and the Harvard Radcliffe Society of Physics Students. An example of such awards is the one given in 2009 for a brassiere that can quickly convert into a pair of protective face masks. It is called the RAD Emergency Bra and described as a brassiere convertible into face masks with a radiation sensor incorporated into the body of the bra in the event of an emergency in which the sensor (located under the front clasps of the brassiere) will change color. (RAD stands for radiation absorbed dose.) To determine the number of RAD units the sensor can be matched with colors listed on the calibration bars found are the service strip attached to the side of the bra. The device is used to warn the wearer of the presence of ionizing radiation in the event of a "dirty" bomb explosion or any other type of nuclear release involving high energy gamma rays.

The first round of Ig Nobel prizes were handed out in 1991, honoring among others, the inventor of the anti-flatulence pill, Beano. The awards have continued every year since then, recognizing inventions such as the Vegomatic, karaoke, and Underease, the world's only airtight, charcoal filtered underwear.

If you are wondering how achievements might attract the attention of Improbable Research, well anyone can send in a nomination, and you can even nominate yourself. The magazine receives more than 5000 new nominations every year, which are added to the collected nominees from previous years. Then the Ig Nobel Board of Governors narrows down a list of finalists in 10 categories; nutrition, peace, archeology, biology, medicine, cognitive science, economics, physics, chemistry and literature. After each finalist has been investigated for authenticity, the board members cast their votes. The awards celebrate "achievements that cannot or should not be reproduced." Even if you are not interested in chemistry, the 2007 award given for turning cow manure into vanilla flavoring should pique anyone's scientific curiosity.

The 2018 Ig Nobel prize for medicine was given to Marc Mitchell and David Warlinger for using roller coaster rides to try to hasten the passage of kidney stones. The anthropology award went to three researchers who collected evidence in a zoo, that chimpanzees imitate humans about as often, and about as well, as humans imitate chimps. The biology prize went to eight individuals who demonstrated that wine experts can reliably identify, by smell, the presence of a single fly in a glass of wine. One person won the nutritional award for calculating that the caloric intake from a cannibalian diet is significantly lower than the caloric intake from most other traditional meat diets. The peace prize was particularly relevant, it was given to scientists who measured the frequency, motivation and effects of shouting and cursing while driving an automobile.

An example of a study that won a prize was conducted at Cornell university where David Dunning and Justin Kruger supplied scientific evidence that incompetence is bliss, for the incompetent person. They staged a series of experiments, involving several groups of people. Beforehand, they made some predictions, most notably that: incompetent people dramatically overestimate their ability; and incompetent people are not good at recognizing incompetence—their own or anyone else's. Their work became the basis for the Dunning Kruger Effect, it occurs when people fail to adequately assess their level of competence—or incompetence—at a task and consider themselves to be more competent than they actually are. This theory is also commonly known as "Mount Stupid", the place where you have enough knowledge of a subject to be vocal about it, without the wisdom to gather the full facts or read around the topic.

It should be mentioned that Ig Nobel prizes are not designed to ridicule science, to the contrary. Prizes honor achievements that make people laugh and then think. Good achievements can also be odd, funny, and even absurd. So can bad achievement, a lot of science gets attacked because of its absurdity, a lot of bad science gets revered despite its absurdity.

ANTS

—*The real masters of our planet.*

ACCORDING TO A recent news report, fire ants were among the creepiest images to emerge from the flood in Houston caused by Hurricane Harvey. Instinctively, they rose up from their underground tunnel systems and literally stuck together to survive by linking their six claws (legs) and clinging to one another in massive rafts and balls that floated and spun in the current. Linking their bodies in a specified manner is evidence of their remarkable cooperation and memory. Fortunately for the ants they possess a coating on their armor- like bodies that repels water. Unfortunately, anyone in contact with the resultant clump of ants can be bitten. The ants inject a stinging venom that burns and develops into a fluid filled pustule and possible death due to an allergenic response.

Ants have best been described many years ago by Dr. Lewis Thomas, a much quoted medical writer. According to Dr. Thomas: "Ants are so much like human beings as to be an embarrassment. They farm fungi, raise aphids as live stock, launch armies into war, use chemical sprays to alarm and confuse enemies, and even capture slaves. And they exchange information ceaselessly—in fact, they do almost everything but watch television."

Ants are indeed remarkable, they have been around since a wasp first shed its wings eons ago. All of its descendants live in the earth, not the ancestral air of insects that fly. Ants are extremely adaptable and successful and can be found in every environment except for the liquid, molten or frozen places of Earth. They can constitute up to 15 percent of the total animal biomass of a tropical rainforest, in the Amazon the combined weight of ants is said to be four times larger than that of other insects in the same area. The number of ants alive at any given time has been estimated conservatively at 100 million billion. If this estimate is correct, and given that each human weighs on average very roughly 1 or 2 million times as much as a typical ant, then ants and people have the same global biomass.

According to convention, ants are classified in the order *Hymenoptera,* as are wasps, termites and bees, they arose during the Cretaceous period, and their history thereafter spanned well over 100 million years to the present time. (Compared to ants, humans are newcomers to this planet and we all can learn a great deal from the ant's highly successful years of experience.) Ants are one of the most successful groups of insects in the animal kingdom because they are social and form highly organized colonies or nests which sometimes consist of millions of individuals. Colonies of invasive ant species will sometimes work together and form super colonies. Fire ants are a great example. There are likely more than 14,000 known ant species, most of which reside in hot climates.

Like no other, ants have become the insect geniuses of chemical communication. Their bodies are crowded with more than forty exocrine glands employed to manufacture pheromones, the chemical compounds used as signals. Ants have enhanced the chemical channel in several ways, variously by mixing pheromones from multiple glands, by giving separate meanings to different concentrations of the same pheromone, and by changing meanings according to context. They have simultaneously added auxiliary signals of touch and vibration conducted through the ant's antennae. The antennae serve as a communication device.

Fire ants have another surprising quality, they can act like a fluid or a solid, depending on the situation. It is the first time this duality has been observed in living things. When under stress these ants form a thick liquid which flows into cracks to quickly repair breaks in any structural damage. Research into this phenomenon could have practical applications as there is great interest in materials that can automatically repair cracks.

Just recently invasive "crazy ants" have began to rapidly displace fire ants in areas across the southeastern United States by secreting a compound that neutralizes fire ant venom, It is the first known example of an insect with the ability to detoxify another insect's venom. The compound has been found to be formic acid which is secreted from a specialized gland at the tip of its abdomen, which is then transferred to the mouth and subsequently smeared on the crazy ant's body.

There is still much to be learned from ants. They live within highly structured and organized cooperatives a system that has been validated through 100 million years of successful experience. Despite crowding, they, unlike honey bees, have managed to avoid high incidences of contagious diseases. Ants have also learned to carry out their tasks for the good of the colony—the ultimate unselfish existence. Some might say that ants are the real masters of our planet.

Bacteriophage therapy

—*Getting closer?*

LONG BEFORE THERE were antibiotics, researchers envisioned using viruses to seek out and destroy bacteria. Now, as organisms continue to develop resistance to existing antibiotics, these viruses, called bacteriophages, are finding new advocates. (The name bacteriophages is derived from the ancient Greek "*phagein*" to devour.) Their discovery may turn out to be one of the great medical findings of the past century. According to the Food and Drug Administration (FDA), bacteriophages (phages) are defined as RNA or DNA viruses that infect bacteria without infecting mammalian or plant cells. Phages are present throughout the environment, and humans are routinely exposed to them at high levels through food and water without adverse effects.

Phages have become a focus of research in the battle against antibiotic resistance in the United States. Although not currently permitted here, phages are used in other countries as antibiotic therapy. Depending on the results of the clinical research, phages may eventually become the treatment of choice for cases where antibiotics fail, but there are a host of problems to be solved before that happens. In February this year the FDA announced the first U.S. clinical trial of an intravenously administered bacteriophage therapy had been approved. The proposed phase 1 and 2 trial will evaluate the safety, tolerability, and efficacy of an experimental treatment for patients with ventricular assist devices who have developed *Staphylococcus aureus* infections. There had been a high unmet need in such patients which are typically very difficult to eradicate with conventional antibiotics.

Most phages have double stranded DNAs encapsulated into an icosahedral shell of protein attached to a tail. At the end of the tail there are proteins that attach to cells. A simple explanation is as follows: the virus particle with its protein and DNA first lands on the outside of the specific bacterial cell and injects its DNA into the cell. The DNA of the bacterial virus then takes over the cell and is converted into a virus factory. The bacterial cell dies and hundreds of virus particles are released. The process is much more complex consisting of a cascade of events involving several structural and regulatory genes.

ADVANTAGES AND DISADVANTAGES OF PHAGE THERAPY

The discovery of viruses that can infect and destroy bacteria was greeted with considerable optimism in the early 1900s. Despite the efforts of a number of investigators, their use was generally abandoned soon after the introduction of antibiotics in the 1940s. Lytic phages, of course, are similar to antibiotics in that they have remarkable antibacterial activity and their theoretical advantages are good reasons for renewed interest. Earlier reported results using phages may have been even better if it had been recognized that there are many types of phages and that each is specific for a special host range of bacteria. This misconception resulted in applications of phage growing on one bacterial host but with little, if any, ability to influence clinical infections caused by other bacterial strains.

Phages have a number of advantages compared to antibiotics. For one, phages are very specific, they usually affect only the targeted bacterial species. Antibiotics target both pathogenic microorganisms and normal microflora. This affects the microbial balance in the patient which may lead to a serious secondary infection. Second, phages replicate at the site of infection and available where they are most needed. Antibiotics are metabolized, and eliminated from the body, and do not concentrate at the site of infection. Third, phages are found throughout nature and it is easy to find new phages when bacteria become resistant to them. This means that selecting new phages is a relatively rapid process that can be accomplished in days or weeks, whereas developing a new antibiotic is a time consuming process that can take several years. Phages appear to be safe as no serious side effects have been described.

Phage therapy is not without disadvantages. There are no internationally recognized studies that attest to the efficacy of phages in human patients. There are a number of publications on phage therapy, but very few papers in which the pharmacokinetics of therapeutic phage preparations is described. Additional research would be needed to obtain the type of pharmacological and toxicological data required by the Food and Drug Administration.

There is a paucity of appropriately conducted, placebo-controlled studies. Because of the high specificity of phages, many negative results may have been obtained because of failure to select phages for the targeted bacterial species. Another concern regarding the therapeutic use of phages is that the development of resistance may hamper their effectiveness. Because bacteria are under constant threat of infection by phages, there are strong selective pressures to acquire resistance.

The Future

The potential for treating infectious diseases with phages has been pursued since their discovery, but for the reasons outlined above, phage therapy is not yet accepted in Western medicine. There still remain many important questions to address before phages can be endorsed for therapeutic use.

Bed bugs

—*Creatures of the night!*

For those of us who travel frequently, there is another major hazard to consider—-bed bugs! Once nearly eradicated, these blood suckers have made a rapid comeback in the United States, Canada, Europe and Australia. This spread may be a global pattern. Traveling to and staying in hotels overnight in New York City is especially problematic as this city appears to be particularly affected. New York, however, is not unique. Bed bug infestations have been reported increasingly in homes, apartments, hotel rooms, hospitals, and dormitories in the United States since 1980. Reports of bed bug infestations in San Francisco doubled between 2004 and 2006. Bed bugs have superseded termites as the number one urban pest and they have evolved to outsmart the latest generation of chemicals used to control them. Bed bugs are no longer kept in check by insecticides called pyrethroids.

Bed bugs, known in Latin as *Cimex* (a bug) *lectularius* (couch or bed), are flightless, nocturnal, blood feeding parasites that preferentially dine on humans. Adult bed bugs are oval shaped, flat, approximately 5 mm long. They resemble unfed ticks or small cockroaches, and are easily visible, even to the untrained eye. (About the size of an apple seed.) Adults are reddish brown in color, whereas immature bed bugs are much smaller and may be light yellow. They have a pyramid shaped head with prominent compound eyes, slender antennae, and a long proboscis tucked underneath the head and thorax. There are 91 species included in the insect family Cimicidae, but only two species, *Cimex lectularius* (human bed bug) *and* lesser known *C. hemipterus* (tropical bed bug), readily feed on humans. Bed bugs feast on sleeping persons every 5 to 10 days and hide during the day in cracks and crevices of beds, box springs, headboards and bed frames. They are apparent only if a special search is made at night.

The common bed bug is one of the most widely recognized insects in the world. Its close association with humans has been documented for more than 4 millennia. The association of *C. lectularius* and humans dates back to 1350 B.C. or earlier, as evidenced by well-preserved bed bug remains recovered in Egypt. According to one researcher, bed bugs are our oldest roommates. Bed bugs are not native to North America, but were introduced by the early colonists in the 17th century. As mentioned earlier, the resurgence of the parasite has been recorded across the globe with an estimated 100-500 percent annual increase in bed bug populations.

Despite the long association between bed bugs and humans, there was little knowledge about the biology of this important insect and its close relatives until recently. This has changed with the increase in global bed bug infestations. Bed bugs have been found to release substantial amounts of histamine that cause an immune response resulting in a rash or respiratory problems. There is also new information regarding aspects of reproduction and ecology. Feeding behavior in *C. lectularius* coincides with periods of minimal host activity, when bed bugs leave their refuge to feed. Bed bugs use two feeding tubes, one to inject an anticoagulant and mild anesthetic, and other to draw blood. An adult fully engorges in 10 to 20 minutes, after which it returns to its hiding place. Both sexes feed only on blood. Egg production in adult females and sperm production in males requires regular blood meals——an essential prerequisite for mating.

Recent studies show that bed bugs around the world have developed a resistance to the chemicals used to control them. Laboratory tests in the U.S., Europe and Africa demonstrate that today's bedbugs can survive pesticide levels a thousand times greater than the lethal dose just a decade ago. Virtually all of these pests have evolved to outsmart the latest generation of pyrethroid insecticides used to control them.

Bed bugs today appear (1) to have nerve cells better able to withstand the chemical effects, (2) higher levels of enzymes that detoxify the lethal substances, and (3) thicker shells that can block insecticides. One possibility to explain the resistance is that many tropical countries have drastically scaled up mosquito control by spraying indoors and providing bed nets impregnated with pyrethroids. This has increased the likelihood of resistance and global travel and trade could have introduced the already-resistant bugs to the United States. Another factor involved in the increased populations is in-breeding. A single female that has been mated is able to colonize and start a new infestation. Her progeny and brothers and sisters can then mate with each other, exponentially expanding the population.

One way to prevent bed bug bites is to cover up. Because bed bugs do not tend to burrow under clothing, it may be possible to avoid bites by wearing pajamas that cover as much skin as possible. It is also wise to place luggage on tables and dressers instead of on the floor. To be completely safe, it is wise to check mattress seams in hotels for bed bug excrement. If your home is infested your best bet is to hire a professional exterminator. They use portable devices to produce steam, heat or freezing temperature to kill the parasites.

BEES

—*Badly needed but vanishing*

A SPECIES OF bee never seen before in Indiana has been discovered at Burnett Woods Nature Preserve in Avon. The bee identified as an Andrena uvulariae, comes amid a decline in pollinators worldwide. Bees began mysteriously vanishing in 2006, first in North America, and then Europe. According to Thor Hanson in his recent book, *Buzz: The Nature and Necessity of Bees,* seemingly healthy bees would simply fly away and never come back. Scientists dubbed the phenomenon, "colony collapse disorder" and launched a massive research effort. So far, little progress has been made and bees continue to die even though collapse disorder peaked quickly and has been on the decline. Experts believe that multiple factors are at play. The main suspects are referred to as the four Ps: parasites, poor nutrition, pathogens and pesticides. Parasites include tiny mites that feed on bee's body fluids. Poor nutrition reflects a widespread loss of flowers in rural landscapes because of the rise of industrial farming practices. The list of pathogens carried by bees includes fungal infections and wing-deforming viruses, many of which get moved around with international trade. Pesticides have received the most attention, and the European Union cited bee concerns in their recent ban of a popular class known as neonicotinoids that are widely used on crops, lawns, gardens and forests. Environmentalists blame climate change as well. It affects the bees in several different ways, including increased temperature and precipitation extremes, increased drought, early snow melt and late harvest events.

There are roughly 20,000 different species of bees and several American bee species have been classified as endangered. In the span of just seven years, the United States saw an approximate average rate of 1.6 million bees die off each year. Bumble bees are an example, their population has plunged almost 90 percent since the 1990s; the species is the key pollinator of blueberries, tomatoes and wildflowers.

Bees greatly contribute to pollinating agricultural crops. It is estimated that they and other insects pollinate almost 90 percent of flowering plants and 70 percent of the 124 world's main crops. Without bees these numbers would begin to rapidly decline. According to the US Department of Agriculture, honey bee pollination alone is valued at more than $14 billion annually. Fortunately, honey bees are not currently endangered.hh

In symbolism and in daily life, the value of bees to people lies rooted in their biology. The modern bee is a marvel of engineering, with wraparound ultraviolet vision; flexible, interlocking wings; and a pair of hypersensitive antennae capable of sniffing out everything from rose blossoms to bombs to cancer. Bees evolved alongside the flowering plants, and their most remarkable traits all developed in the context of that relationship.

Flowers provide bees with the ingredients for honey and wax as well as the impetus for navigation, communication, cooperation, and, in some cases, buzzing itself. In return, bees perform what is their most fundamental and essential service. Yet, oddly, it's one that people didn't begin to understand—let alone appreciate—until the seventeenth century with discoveries by the great Dutch savant, Swammerdam. He founded the true methods of scientific investigation; helped to invent the microscope, contrived injections to ward off decay, was the first to dissect the bees, and by the discovery of the ovaries and the oviduct definitely fixed the sex of the queen, hitherto looked upon as a king, and threw the whole political scheme of the hive into most unexpected light by basing it upon maternity. Finally he produced woodcuts and engravings so perfect that to this day they serve to illustrate many books on apiculture (bee keeping).

One of the best books about bees was written long ago by Maurice Maeterlinck and published in 1901. In *The Life of the Bee,* he describes them as follows: "They are the soul of the summer, the clock whose dial records the moments of plenty; they are the untiring wing on which delicate perfumes float; the guide of the quivering light-ray, the song of the slumberous, languid air; and their flight is the token, the sure and melodious note, of all the myriad fragile joys that are born in the heat and dwell in the sunshine. They teach us to tune our ear to the softest, most intimate whisper of these good, natural hours. To him who has known them and loved them, a summer where there are no bees becomes as sad and as empty as one without flowers or birds."

According to the aforementioned Thor Hanson, bees helped shape the natural world where our own species evolved, and their story often comingles with our own. To understand them, and ultimately to help them, we should appreciate not only where bees came from and how they work, but also why they've become one of the only insects to inspire more fondness than fear.

BILL BRYSON

―◆―

—*A writer worth reading*

'HUMANS BELONG TO the group of conscious beings that are carbon-based, solar system-dependent, limited in knowledge, prone to error and mortal.'

If someone were to ask me who my favorite contemporary writer is, I would without hesitation say Bill Bryson, of course. Although not thought of as a science author, he has written about the universe, the human body and in his book about Australia, snakes, spiders and deadly fish. He is also prolific, in chronological order, his titles include: *The Lost Continent: Travels in Small Town America, The Mother Tongue: English and How it Got That Way , Neither Here Nor There, Made in America: An informal History of the English Language in the United States, Notes From a Small Island, A Walk in the Woods: Rediscovering America on the Appalachian Trail, Notes From a Big Country, Down Under, Bryson's Dictionary of Troublesome Words, At Home: a Short History of Private Lives, Walk About, A Short History of Just About Everything, Seeing Further: The Story of Science and the Royal Society, The Life and Times of the Thunderbolt Kid, Shakespeare: the World as Stage, One Summer America 1927* and his latest, *The Body:A Guide for Occupants*. Most of the books are classified as either travel, language, history, or science. Moreover, they are all entertaining and fun to read. He won the Aventis Prize for Science Books in 2004, and was awarded the Descartes Science Communication Prize in 2005.

I am currently reading his latest, *The Body: A Guide for Occupants*. Anyone interested in science will find it most enlightening. In the section on elements that make up the human body, Bryson had this to say: "That is unquestionably the most astounding thing about us—that we are just a collection of inert components, the same stuff you would find in a pile of dirt. I've said it before in another book, but I believe it's worth repeating: the only thing special about the elements that make you is that they make you. That is the miracle of life." He also wrote that it takes 7 billion billion billion (that's 7,000,000,000,000,000,000,000,000,000, or 7 octillion) atoms to make you. No one can say why those 7 billion billion billion have such an urgent desire to be you. They are mindless particles, after all, without a single thought or notion between them. Yet somehow for the length of your existence, they will build and maintain all the countless systems and structures necessary to keep you humming, to make you you, to give you form and shape and let you enjoy the rare and supremely agreeable condition known as life. Bryson is truly gifted, he writes sentences you like to read over and over again.

In his book *Seeing Further: The Story of Science and the Royal Society*, Bryson described the background behind Reverend Thomas Bayes and derivation of his famous theorem in the mid 1750s. The remarkable feature is that it had no practical applications in his own lifetime. Although simple cases yield simple sums, most uses demand serious computational power to do the volume of calculations. So in Bayes' day it was simply an interesting but largely pointless exercise. Bayes evidently thought so little of his theorem that he didn't bother to publish it. It was a friend who sent it to the Royal Society in London in 1763, two years after Bayes' death, where it was published in the Society's Philosophical Transactions with the modest title of 'An Essay Towards Solving a Problem in the Doctrine of Chances'. In fact, it was a milestone in the history of mathematics. Bayes' theorem is used routinely in the modelling of climate change and weather forecasting generally, in interpreting radiocarbon dates, in social policy, astrophysics, stock market analysis, clinical trials and wherever else probability is a problem. And its discoverer is remembered today simply because nearly 250 years ago someone at the Royal Society decided it was worth preserving his work, just in case.

Bryson 's research frequently uncovers new and odd information. His book about Shakespeare is surprising in that there is little known about the greatest writer of them all. "Although he left nearly a million words of text, we have just fourteen words in his own hand—his name signed six times and the words "by me" on his will. Not a single note or letter or page of manuscript survives. (Some authorities believe that a section of the play Sir Thomas More, which was never performed, is in Shakespeare's hand, but that is far from certain.) We have no written description of him penned in his own lifetime. The first textual portrait—"he was a handsome, well-shap't man: very good company, and of a very readie and pleasant smooth witt"—was written sixty-four years after his death by a man, John Aubrey, who was born ten years after that death.

Copying nature's ways

Biomimicry is the relatively new engineering skill that takes a leaf from nature's book by building intricate structures with surprising new properties derived from nature's own nanotechnology. Examples include using spider silk to manufacture tendons and ligaments, and using the leaves of the lotus plant to self clean surfaces. There is also the amazing adhesion of a gecko's foot and how these properties can be applied to apparel and wearable prosthetics. The barbs of a porcupine's quills can serve as a basis for development of bio-inspired devices such as tissue adhesives and hypodermic needles. The wonders of nature are truly amazing, perhaps best summarized in the following paragraph from Janine Benyus' new book entitled *Biomimicry*:

"When we stare this deeply into nature's eyes, it takes our breath away, and in a good way, it bursts our bubble. We realize that all our inventions have already appeared in nature in a more elegant form and at a lot less cost to the planet. Our most clever architectural struts and beams are already featured in lily pads and bamboo stems. Our central heating and air-conditioning are bested by the termite tower's steady 86 degrees F. Our most stealthy radar is hard of hearing compared to the bat's multifrequency transmission. And our new "smart materials" can't hold a candle to the dolphin's skin or the butterfly's proboscis. Even the wheel,

which we always took to be a uniquely human creation, has been found in the tiny rotary motor that propels the flagellum of the world's most ancient bacteria. Humbling also are the hordes of organisms casually performing feats we can only dream about. Bioluminescent algae splash chemicals together to light their body lanterns. Arctic fish and frogs freeze solid and then spring to life, having protected their organs from ice damage. Black bears hibernate all winter without poisoning themselves on their urea, while their polar cousins stay active, with a coat of transparent hollow hairs covering their skins like the panes of a greenhouse. Chameleons and cuttlefish hide without moving, changing the pattern of their skin to instantly blend with their surroundings. Bees, turtles, and birds navigate without maps, while whales and penguins dive without scuba gear. How do they do it? How do dragonflies outmaneuver our best helicopters?"

This paragraph clearly demonstrates that nature is the best source of answers to the technological, biological, and design challenges that we face as humans. Scientists have already identified more than two million species of life on earth; some estimate that there may be as many as one hundred million. Each one has evolved hundreds of optimized solutions to life's challenges, many of which can be readily applied to the very problems facing human enterprise and survival. By constantly creating conditions conducive to life, with zero waste and a balanced use of resources, nature is clean, green, and sustainable. Following nature's design mastery, we can achieve greater wealth and economic sustainability. Even the most bizarre ideas are possible. According to Jay Harman in his book *"The Shark's Paintbrush"*, hippopotamus perspiration has unheard of value. It turns out that hippo sweat provides a highly effective, four-in-one sunblock. We humans perspire by allowing salt water to leave our pores, using the physics of evaporation to cool the skin. Hippos—long-lost cousins of whales and dolphins—solve more than just a cooling problem by secreting a blend of chemicals that takes care of many challenges simultaneously. Besides being an excellent, nontoxic sunscreen (though perhaps a little aromatic in its natural form), hippo sweat is insect repellent, antiseptic and antifungal.

One of the best examples of utilizing biomimicry was employed by George DeMestral. He was an inventor trained as an electrical engineer, who lived in Lausanne, Switzerland. His pastime of hunting in the lower slopes of the Jura mountains led to his discovery. In 1948, when de Mestral returned home for one of his walks, his dog and his pants were covered in burs. (Cockleburs are the plant seed sacs that cling to animal fur to travel to fertile new planting grounds.) Instead of just picking them off, he marveled at their tenacity and examined the burs microscopically. He noticed all the small sharp hooks that enabled the bur to cling to the tiny loops in the fabric of his pants or animal fur.

This serendipitous occasion led George de Mestral's decision to design a unique, two sided fastener, one side with stiff hooks like the burs and the other side with soft loops like the fabric of his trousers. The product called Velcro (a combination of velour and crochet) came to the market in 1955. The American patent expired in 1978 and George de Mestral died in 1990. Hook and loop fasteners are now a major product world wide and the Velcro Corporation (which George started) is still the major producer. It is a multimillion dollar industry and proof that you can't beat Mother Nature.

98.6 F

—It is not the average body temperature anymore!

MOST EVERYONE BELIEVES that the average body temperature is 37°C (98.6°F). This number has been around a long time and considered canonical. (It follows a principle that is a basic undisputed rule or law.) In truth, each of us has his or her own individual "normal" body temperature, which may be slightly higher or lower. Our bodies also constantly adapt their temperature to environmental conditions. It goes up when we exercise, for instance. And it is lower at night, and higher in the afternoon than in the morning. Strictly speaking, body temperature refers to temperature in the hypothalamus, a small region of the brain located at its base and in the vital internal organs. Because we cannot measure the temperature inside these organs, temperature is taken on parts of the body that are more accessible, i.e., orally, rectally, vaginally, axillarily (under the armpit) or in the ear. Ear and rectal temperatures tend to be a half degree higher than oral temperature. Axillary temperature tends to be one degree lower.

There is a recent study from Stanford University that analyzed 677,423 temperatures collected from 189,338 individuals, it indicated that In the United States, the normal, oral temperature of adults is, on average, lower than the widely accepted 98.6° F (37°C) established in the 19th century by Professor Carl Reinhold August Wunderlich in Germany. Professor Wunderlich pioneering and extensive work was published in 1851 and based on results from an astounding 25,000 patients. Dr. Wunderlich reported that "The average normal temperature of the healthy human body in its interior, or in carefully covered situations on its surface, varies, according to the plan of measurement, from 98-6° to 99*5° F. (37° to 37*5° C). It is about 98*6° in the well-closed axilla, and a few tenths of a degree higher in the rectum and vagina." The Stanford University comprehensive study determined that body temperature has decreased over time. It provides a framework for understanding changes in human health and longevity over the past 157 years. Using measurements from three cohorts—the Union Army Veterans of the Civil War measured in the years 1860–1940, the National Health and Nutrition Examination Survey taken in 1971–1979, and the Stanford Translational Research Integrated Database Environment from 2007–2017—the researchers determined that mean body temperature in men and women, after adjusting for age, height, weight and, in some models date and time of day, has decreased monotonically by 0.03°C per birth decade. A similar decline within the Union Army cohort as between cohorts, makes measurement error an unlikely explanation.

There have also been at least 27 other modern studies reporting mean temperatures uniformly lower than Wunderlich's estimate. Recently, an analysis of more than 35,000 British patients with almost 250,000 temperature measurements, found mean oral temperature to be 36.6°C, confirming this lower value. Remaining unanswered is whether the observed difference between Wunderlich's and modern averages represents true change or bias from either the method of obtaining temperature (axillary by Wunderlich vs. oral today) or the quality of thermometers and their calibration. Wunderlich obtained his measurements in an era when life expectancy was 38 years and untreated chronic infections such as tuberculosis, syphilis, and periodontitis afflicted large proportions of the population. These infectious diseases and other causes of chronic inflammation may well have influenced the 'normal' body temperature of that era. Wunderlich's book On the Temperature in Diseases: A Manual of Medical Thermometry published in 1871 has been translated into English.

The question of whether mean body temperature is changing over time is not merely a matter of idle curiosity. Human body temperature is a crude surrogate for basal metabolic rate which, in turn, has been linked to both longevity (higher metabolic rate, shorter life span) and body size (lower metabolism, greater body mass). The Stanford group speculated that the differences observed in temperature between the 19th century and today are real and that the change over time provides important physiologic clues to alterations in human health and longevity since the Industrial Revolution.

FINAL THOUGHTS

According to a recent Wall Street Journal article the Stanford study could offer a clue about other physiological changes that have occurred over time. The article also noted that the accumulation of evidence on body temperature should change medical norms, guidelines and thresholds about when and how to treat patients. The article does note, however, that a thermometer reading matters less than how we feel. "If you are sick, you're sick, regardless of your temperature."

THE BREATH OF LIFE

SEVERAL YEARS AGO a siege of bronchitis caused me to suffer with a severe cough which in turn resulted in sudden shortness of breath, gasping for air, and likely my demise. Fortunately, a trip to the emergency room, a Medrol dosepack, and two inhalers solved my problem. Since then I no longer take the simple act of breathing for granted and have developed an acute interest in anything related to the process of respiration and the associated anatomy. My research led to a number of facts I was unaware of, or had forgotten. For example, I learned that absence of breath is the major way to certify or determine that someone is dead. In Renaissance times a feather would be placed on the lips to see if air was moving in and out of the lungs.

A human breathes about 20 times per minute, taking in 13 pints of air during that time. Breathing brings air (oxygen, nitrogen and traces of carbon dioxide) into the blood which circulates it throughout the body. Air then passes through the larynx and trachea, which it is directed to the chest cavity. In the chest, the trachea splits into two bronchi, which lead to the lungs. Each lung consists of more than 100,000 small airways, more than 200 million alveoli, and has a surface area greater than of a tennis court. Oxygen passes into the alveoli and diffuses through capillaries into the blood stream where red blood cells carry it to all parts of the body. At the same time blood from the veins release carbon dioxide into the alveoli, and it is expelled. Everything inhalable from tobacco smoke to indoor and outdoor air pollutants, vapors to medications directly delivered to the lungs, has an effect on the both the lung itself, and, if entering the bloodstream, on other organs.

All animals require oxygen and must rid the body of carbon dioxide. Breathing through the nose has a number of advantages, it acts as a filter and retains particles in the air, such as pollen. The nose also adds moisture to the air to prevent dryness in the lungs and bronchial tubes and it warms cold air to body temperature before it reaches the lungs. I was not aware that the nose produces nitric oxide, which improves the lung's ability to absorb oxygen. Nitric oxide enhances the ability to transport oxygen throughout the body, including inside the heart and relaxes vascular smooth muscle which allows blood vessels to dilate. Moreover, nitric oxide has antifungal, antiviral, antiparastic, and antibacterial properties and helps the immune system fight off infections.

The average person breathes around thirty thousand times during a twenty-four hour period. As a newborn we draw our first breaths automatically, perhaps forty times per minute, and it slows down to closer to half that rate as we age. Fortunately we breathe without thinking about it. We are hardwired to continue this labor reflexively, even in our sleep.

The effort of breathing consumes roughly 3 percent of our metabolic energy at rest, all in order to pull the equivalent volume of air the size of a grapefruit into our lungs. This process traps trillions of air molecules within our chests like fish in a net. Only a few of them, the oxygens, are what we are after. An average adult uses nearly two pounds of them every day, and this particular breath full will help to keep us alive for the next few minutes. Your lungs are not necessarily the only route that oxygen takes to your blood, however. You also breathe a little through your eyes. So vital are these oxygen particles that the cells in the transparent surfaces of your eyes absorb them directly from the atmosphere to supplement the meager supply that your blood vessels send to them, as do many of the cells of your skin.

Unlike eating or drinking, you have to inhale oxygen continuously because, apart from the inflatable bags of your lungs, you can't safely hold much of it inside of you. Even if you could purify and compress a lot of it into some internal storage space, you wouldn't want to do so. Left unguarded within your body, it can attack and damage your cells, and it is toxic in high doses. According to Curt Stager, in his book entitled *Your Atomic Self: The Invisible Elements that Connect You to Everything Else in the Universe*, "Oxygen levels must be consumed in controlled sips, using it immediately and efficiently, and then take more from the sea of atoms that surrounds you."

Cancer

—Prevention may be the only solution!

Noted oncologist Azra Raza in her new book, *The First Cell*, describes the reasons why cancer is a dreaded and often fatal illness. According to her extensive experience she wrote as follows: "Treating cancer as one disease is like treating Africa as one country. Even in the same patient, it is not the same disease at two sites or at two different points in time. Vicious and self-obsessed, it learns to grow faster and become stronger, smarter, and more dangerous with each successive division. It is a perfect example of intelligence at a molecular level, able to perceive its environment and take actions that maximize its chances of survival. A feedback loop, using past performance to improve its efficiency, forms the basis of its seemingly purposeful behavior. It learns to divide more vigorously with time, invading new spaces, mutating to turn the expression of pertinent genes off and on, enhancing its fitness to the landscape, optimizing seed-soil cooperation." In her opinion today's practices have done little to improve treatment despite limited success and all of the millions of research papers claiming success.

In her career, Dr. Raza has cared for hundreds of patients and unfortunately, has little regard for significant future progress in cancer research. Until the day a breakthrough occurs, if it ever does, then the best course of action is prevention. We do know that certain life style changes can diminish the chance of contracting cancer, this includes not smoking, the reason s that using any type of tobacco can place you on a collision course with the disease. It is wise to eat a healthy diet, maintain a healthy weight and be physically active. Other suggestions are to protect yourself from the sun (to avoid skin cancer), and to become vaccinated. (The Human Papillomavirus (HPV) vaccine acts against genital warts and/or different types of HPV viruses that can cause cancer.) Other suggestions are to avoid risky behaviors (contact with certain chemicals, including radon) and to obtain regular medical care.

Radon

Preventing the possibility of allowing radon, a silent radioactive gas from seeping into your home is not mentioned near enough. Radon is an invisible, odorless and tasteless gas, with no immediate health symptoms, that comes from the breakdown of radioactive elements, such as uranium, which are found in different amounts in rock and soil throughout the world. Radon gas in the soil and rock can move into the air and into underground and surface water. According to the American Cancer Society, radon is present outdoors and indoors, and normally found at very low levels in outdoor air and in drinking water from rivers and lakes. It can be found at higher levels in the air in houses and other buildings.

Most exposure to radon comes from being indoors in homes, offices, schools and other buildings. The gas can enter buildings through cracks in floors or walls; reconstruction joints; or gaps in foundations around pipes, wires or pumps. Levels are highest in the basement or crawl space. People who spend much of their time in basements at home or at work have a greater risk of being exposed.

Radon breaks down into so-called solid radioactive elements such as polonium-218, polonium -214 and lead-214. They can attach to dust and other particles and breathed into the lungs. As radon elements in the air break down, they give off radiation that can damage the DNA inside body cells and eventually, cancer.

The Surgeon General of the United States stated that, "Indoor radon is the second leading cause of lung cancer in the United States and breathing it over a prolonged period can present a significant health risk." More than 200,000 Americans die from radon-associated lung cancer every year. Millions of homes have an elevated radon level, Kosciusko county is no exception. The levels of the deadly radon gas in homes and other buildings depend on the characteristics of the rock and soil in the area. Elevated levels vary greatly in different parts of the United States, sometimes within neighborhoods. **Fortunately, radon can be detected with a simple test and fixed through well-established venting techniques**. Recent studies reveal that radon gas levels can fluctuate wildly day to day and that short-term tests can give a false sense of alarm or a false sense of security as they cannot precisely predict long-term exposure. The only reliable way to measure levels is with a long term testing kit, 90 days or more.

Final thoughts

One way to learn more about cancer control including prevention, early detection, treatment and survivorship is by visiting the Kosciusko County Cancer Consortium website: https://livewellkosciusko.org/cancer-consortium. You can easily sign up for information worth learning about.

CANDIDA AURIS

—A mysterious and fatal yeast

I HAVE WRITTEN a number of columns about the threat of bacterial resistance to antibiotics and how it has affected public health worldwide. Not enough has been done to reduce the problem and few new antibiotics have been developed and approved for sale. Those that are available may be unsupported because of their cost. There is another danger lurking as well, it concerns resistance to yeast or fungi. (Yeasts are a type of fungi.) One example is *Candida albicans*, that preys on people with weakened immune systems, and it is quietly spreading and virulently affecting severely ill patients in nursing homes in the United States and across the globe. Yeasts and fungi are much different than bacteria. They are multicellular, eukaryotic organisms, while **bacteria** are single-celled prokaryotes. The cells of **fungi and yeasts** have nuclei that contain the chromosomes and other organelles, such as mitochondria and ribosomes. **Bacteria** are much smaller than either yeast or **fungi**, do not have nuclei or other organelles and cannot reproduce sexually.

David Gruby, a Hungarian physician, is generally recognized as the discoverer of fungi that produce diseases. He built an excellent microscope in the early 1840s and practicing physicians have attained expertise in as many specialized fields.

Among other gifts, Sherrington was also an outstanding teacher, his students include Nobel laureates Ragnar Granit, Sir John Eccles and Howard Florey, as well as the American pioneer in brain surgery, Harvey Williams Cushing. In 1905 he was awarded the Royal Medal, in 1922, he became a Knight Grand Cross of the Order of the British Empire and in 1924 was awarded the Order of Merit. At the end of the first World War, his textbook *Mammalian Physiology: a Course of Practical Exercises* for medical students was published. In 1925 he produced a wartime poem collection which was widely praised. He was awarded the Copley Medal in 1927. Sherrington received the Nobel Prize in 1932 for discovering the different functions of nerve cells, sharing the award with Edgar Douglas Adrian. In 1940, when he was eighty three, his philosophy book, Man on His Nature was published. The book and its revised edition in 1951 described work done by Jean Fernel, a 16th century French physician and explored philosophical thoughts about the mind, nerve cells, human existence, and modern theology. (Fernel taught philosophy and introduced the term "physiology" to characterize the body's function and was the first person to describe the spinal canal.) The book went through several editions and was voted one of the hundred best books of modern Britain in 1951.

EARLY HISTORY

Sherrington was born on November 27, 1857 in London but grew up in East Anglia. He was heavily

CHARLES SHERRINGTON (1857-1952) ONE OF A KIND.

I HAVE PREVIOUSLY written about a number of outstanding doctors and research scientists, including James Parkinson, David Nachmansohn, Emory Rovenstine, John Sulston and Otto Warburg, but I overlooked Charles Sherrington. He may be the most important of them all. His contributions to medical research were astounding and most people (like me) have no idea who he is or was. I first read about him in Bill Bryson's new book, *The Body*, and according to the author, Charles Sherrington was one of the greatest and most inexplicably forgotten British scientists of the modern era. He was lifted straight out of a nineteenth century boy's adventure story. Sir Sherrington was a brilliant student who did seminal work on tetanus, industrial fatigue, diphtheria, cholera, bacteriology and hematology. He proposed the reciprocal process for muscles, which states that when one muscle contracts, a companion muscle must relax. His textbook, *The Integrative Action of the Nervous System*, is one of the three landmark books written about medicine, the others are Newton's *Principia* and Harvey's *On the Motion of the Heart*. The book shaped an understanding of the central nervous system. Sherrington has been described as an English neurophysiologist, histologist, bacteriologist and a pathologist. Few if any began a study of the pathology of body fluids. His attempt at microscopic differentiation of pus from other pathological substances, like mucus or sputum, was a careful, original investigation in a new field of medicine. In 1841, Dr. Gruby subsequently found a fungus in favus, a contagious skin disease situated in hair follicles, and shortly thereafter discovered *C. albicans (then Oidium albicans)* the cause of thrush in children. His work was recognized in the New England Journal's 200th Anniversary issue. In 1923 Christine Marie Berkhout named what we now call *Candida albicans,* for the white robe, *toga candida,* worn by Roman senators and senatorial candidates. Albicans also comes from Latin, *albicare* meaning "to whiten." A number of Candida species have been discovered through the years.

Now in addition to *C. Albicans* and others, there is new fungus to worry about and it could be even more deadly. The organism, C. auris (the latin name for ear), is a species reported in Asia as a rare cause of ear infections in 2009; it had not been found among large repositories of samples collected prior to 2013. Retrospective review of Candida strain collections found that the earliest known strain of *C. auris* dates back to 1996 in South Korea.) Its virulence is particularly life threatening because bacterial resistance to antibiotics now includes a worldwide emergence of similar development for drugs used to treat fungal diseases. The inability to control these infections presents a grave risk for both human and animal health. Like *C. albicans, C. auris* preys on people with weakened immune systems, and it is quietly spreading across the globe. According to the New York Times, over the past five years it was found carbon atoms have been dispersed, scattered far and wide throughout today's Universe in cosmic dust and gas. But vast numbers of those very first atoms of carbon have become intermingled with our modern world, indistinguishable from atoms formed in much later events.

Stars are the engines of chemical evolution. Subjected to the unimaginable heat and pressure of stellar interiors, hydrogen fused into helium, while triplets of helium nuclei fused into carbon—a slow process, to be sure, but stars are around a long time. And so, carbon gradually increased in concentration, ultimately to become the fourth-most-abundant element in the Universe, with almost 5 carbon atoms for every 1,000 hydrogen atoms.

Your body contains more than 100 trillion trillion atoms of carbon, about 18 percent of its content. It follows inevitably that trillions of those atoms must be the very same carbon nuclei that formed so long ago in the throes of the Big Bang Nucleosynthesis—atoms inseparable from the more recent hoard of carbon forged in stars. And the same is true of your essential oxygen atoms and your nitrogen atoms, not to mention all that primordial hydrogen—other elements essential to life.

Final thoughts

Chemically, carbon is described as a non-metal solid that occurs in three completely different structural forms (diamond, graphite, and fullerenes). Fullerene is the form of carbon in which the atoms are arranged in soccer-ball shapes. Carbon has been known since ancient times but was first recognized as an element in the second half of the 18th century. Antoine Lavoisier proposed carbon in 1789 from the Latin carbo meaning "charcoal." Carbon is listed as Element 6 in the Periodic Table.

Charles Krauthammer

—Physician, historian, commentator—a great loss!

I watch television news now much less than I did during the years that Charles Krauthammer appeared on Fox News. His commentaries were original, informative, often humorous, analytical, and always timely. The topics were broad. He was comfortable discussing current events, politics, science, medicine, history, philosophy, sports (particularly baseball), space flight or logic. His death last year left us without a replacement and we are a saddened and a less educated nation because of it. Fortunately, there are two books available containing his columns and speeches and they are a delight. The first is *Things That Matter,* and the most recent compiled by his son, *The Point of it All.* I read some of the chapters over and over again and try to learn from at least one of them at bedtime. With all that is going on in politics today, I would relish listening to or reading his comments.

Before he died, Krauthammer said that "I am leaving this life with no regrets. It was a wonderful life—full and complete with the great loves and great endeavors that made it worth living. I am sad to leave, but I leave with the knowledge that I lived the life that I intended."

Charles was born on March 13, 1950 and died on June 21, 2018. He graduated from McGill University in Canada, attended Balliol College, Oxford and received his medical degree from Harvard. Krauthammer spent seven years in Boston, four as a student, and three as a psychiatric resident at Massachusetts General Hospital. While at Harvard at age 22, he suffered a diving accident that severed his spinal cord and left much of his body paralyzed. He did not hide his disability, noting that "all it means is whatever I do is a little bit harder and probably a bit slower with a little more effort."

As a historian, Krauthammer was fond of Winston Churchill, finding him indispensible. Without Churchill, Krauthammer wrote, the world would be unrecognizable, impoverished and tortured. Nazism would have prevailed and civilization would have descended into a darkness the likes of which it have never known. According to Krauthammer, Churchill single handedly saved Western civilization and brought victory to the war. He then immediately rose to warn prophetically against its sister barbarism, Soviet communism.

Krauthammer was equally enamored with scholars who love math and he wrote about them. One, Paul Erdos, a prodigiously gifted and productive mathematician was a favorite. Erdo's whole life was so improbable that no novelist could have invented him. Erdos had no home, no family, no possessions, no address. He went from conference to conference, university to university, knocking on the doors of mathematicians throughout the world, and moving in. When Erdos died, he left no survivors, but in reality he did leave hundreds of scientific collaborators, and 1500 mathematical papers produced with them. (This is an astonishing legacy in a field where a lifetime product of 50 papers is considered extraordinary.)

As a scientist, Krauthammer asked in one of his columns whether we are alone in the universe. The discussion included reasons why there has been no evidence of such life. According to Fermi, the great physicist, "All of our logic, all our anti-isocentrism, assures us that we are not unique—that they must be there. And yet we do not see them." One answer may be that advanced societies have destroyed themselves. Krauthammer remarked that we too are near extinction, with nuclear weapons in the hands of half-mad tyrants (North Korea) and possibly radical zealots like Iran. He relies on politics to restrain such folly. He summarized his thoughts by noting that politics—in all its grubby, grasping, corrupt, contemptible manifestations is sovereign in human affairs and that everything rests upon it.

There were few limits to Krauthammer's commentaries. He was a strong believer in the English language as a requirement for immigrants and its means for assimilation. Krauthammer felt that making English an "official" language would make clear what is expected for those who wish to live here. Every citizen upon entering America, and its voting booth, should minimally be able to identify the words *President* and *Vice President* and *county commissioner* and *judge*. According to Krauthammer, those who come to this country, swear allegiance and accept its bounty, must understand how to join its civic culture, in English.

Final thoughts

In the comments written in the back of *Things that Matter,* one columnist described Krauthammer as possessing a steel-trap logic, epic wit, and a profound sense of history. He was surely our best communicator, the composer of America's greatest prose. He will be missed.

influenced by his stepfather, a physician with a strong interest in the arts. After his schooling, Sherrington began medical studies at St. Thomas' Hospital in London but then transferred to Cambridge, where he was to gain first-class honors. While still a student, he attended the International Medical Congress in London and was especially interested in David Ferrier's evidence for the presence of a motor area in the monkey brain. Next came a period of study in Germany with the noted neuropathologist, Rudolf Virchow and bacteriologist Robert Koch, followed by a return to London and a lectureship at St. Thomas' Hospital Medical School.

Sherrington was appointed professor of physiology at University College, Liverpool, in 1885. It was in Liverpool that Sherrington did much of his best work, including the study with Albert Leyton in which the details of the cortical motor representation were mapped out in several primate species. By now famous, Sherrington was sought after, and visited by, clinicians and scientists working on the nervous system. His reputation was consolidated by an invitation to become the Waynflete Professor of Physiology at Oxford, a post he assumed in 1914. From that time onward, Sherrington concentrated his laboratory work on the exploration of the spinal cord, using the spinal reflexes to investigate central excitation and inhibition. During this period, which continued until his retirement in his 70s, Sherrington continued to teach. As described by Alan McComas in his book, Sherrington's Loom, "Sherrington had a long life, dying in a retirement home at 94; right until the end, however, this small man, the recipient of numerous awards and honorary degrees and the greatest living authority on the

Charles Sherrington (1857-1952) one of a kind.

nervous system, remained in full possession of his faculties and was still able to discuss the workings of the brain with his visitors."

FINAL THOUGHTS

Bill Bryson's book, *The Body*, mentioned earlier, contained a number of favorable comments about Sherrington, perhaps the best is "that he was by all accounts, a wonderful man, devoted husband, gracious host, delightful company, and beloved teacher." An epitaph worth striving for.

CICERO

—An ancient Roman worth emulating!

THIS YEAR MORE than ever I have spent time on the political process, watching debates, congressional actions and learning more about government and its history. Unfortunately, this is another year where I continue to be disenchanted with our representatives, their demeanor, and longevity in office. Their manner of speaking is lacking as well. In my opinion, there is no one that can be called an orator. (Oratory is defined as the art of speaking in public eloquently or effectively.) We had been blessed by men from the past who aroused us by their passionate, well delivered speeches, men like Ronald Reagan, Martin Luther King, John Kennedy and Franklin Roosevelt. (Younger readers can see them on You Tube.) Even though there are more than 300 million people alive today in the United States to choose from, none can compare to those gentlemen from yesteryear. My suggestion for today's leaders is to read more about Cicero, perhaps the greatest orator of all time. His eloquent speeches and his handbooks on the philosophy of speaking are fortunately still available. Marcus Tullius Cicero was born January 3, 106 BC. He was a Roman statesman, lawyer and philosopher who served as consul (the highest office) in the year 63 BC. He came from a wealthy municipal family of the Roman equestrian order. (Equites were members of the Roman cavalry.)

Cicero could not claim to be a native Roman. In fact, he did not want to. He held Roman citizenship and owed Rome his primary loyalty, but his origins lay in a Volscian tribe that had fought for many wars with the fledgling city-state on the Tiber before accepting defeat, assimilation and ultimately full civic rights: "We consider both the place where we were born and the city that has adopted us as our fatherland." His dual nationality is central to an understanding of Cicero's personality. He had that passionate affection for Rome and its traditions which many newcomers feel when they join an exclusive club.

As a student, young Cicero went through the regular curriculum—grammar, rhetoric, and the Greek poets and historians. Like many other youthful geniuses, he wrote a good deal of poetry of his own, which his friends, as was natural, thought very highly of at the time, and of which he himself retained the same good opinion to the end of his life, as would have been natural to few men except Cicero. But his more important studies began after he had assumed the 'white gown' which marked the emergence of the young Roman from boyhood into more responsible life—at sixteen years of age. He then entered on a special education for the bar.

Cicero is known for his rhetorical skills and published methods. Speakers have been using his suggestions for 2000 years. His focus during debates was on the emotions of the listener—not science or truth. He wrote "Nothing is more important than emotion." Repetition was important as well—he believed in repeating points over and over again. He was a firm advocate of vivid language and using words that are simple and clear. Cicero did not hesitate in exaggerating, or accusing an opponent as being guilty by association, often without actual evidence. He often conjured up accusations without truth and thus used fear mongering as a strategy. Cicero was not remiss describing his opponent's personal defects, appearance or associations. Using catch phrases was another ploy and while many do not survive English translation, one of interest is the Latin phrase "*Audaciter territas, humiliter placas*" or " You blusteringly threaten—- you cringingly beg." Many of today's jingles or aphorisms stem from Cicero's original ideas. He implored divine providence in his speeches while characterizing his opponents as evil and gave testimony without evidence. Oftentimes Cicero used the technique of diversion and distraction to achieve favors all without being concerned with what is morally or ethically correct. Effectiveness was what mattered. He used humor to his advantage and relied on graphic visual aids in his presentations—somewhat akin to today's powerpoint. The key to an effective speech according to Cicero is the delivery, including hand and body gestures—it should be lively and engaging.

The fact that Cicero did not appear to be concerned with what is morally or ethically correct in discourse, his life was quite different, in fact polar opposites. Cicero lived by the principles of honorable behavior based on wisdom, justice, magnanimity and propriety. His treatise "*On Obligations*", has played a seminal role in the formation of ethical values in the western world. It has become the foremost guide to good conduct and has been called the best book on morality and ethics ever written. It should be required reading for politicians of all parties.

Final thoughts

One way to help encourage and improve public speaking is to offer courses on oratory in our high schools, like the Romans did. Becoming a member of Toastmasters is another good suggestion.

Circadian Rhythms

—*Another of nature's wonders.*

A RECENT MEDICAL study found that cuts and burns sustained during the day heal 60 percent faster than those occurring at night. Similarly, wounds that happen between 8 pm and 8 am were classed as 95 percent healed after an average of 28 days, compared with an average of 17 days for daytime wounds. This sparked my interest and I began to investigate other effects the "biological clock" or "circadian rhythms" have on our well being. (I will use both terms interchangeably.) Scientists are increasingly finding new ways circadian rhythms affect us. For example, circadian rhythms can influence hormone release, eating habits, digestion, body temperature, sleep, and even the sense of smell. Disruption in circadian rhythms can lead to the increased incidence of many diseases, including metabolic disease and cancer. My previous experience and knowledge about the subject was limited to "jet lag", a recognized sleep disorder I suffered from traveling to the west coast or to Europe, and more recently, chronotherapeutics, a topic of interest to prescribers, pharmacists and patients.

According to the National Institutes of Health circadian rhythms are physical, mental, and behavior changes that follow a daily cycle. They respond primarily to light and darkness in an organism's environment, and they are found in most living things. Circadian rhythms can be thought of as part of a biological clock, an innate timing device composed of specific proteins that interact with cells throughout the body. The possession of some form of clock permits organisms to optimize physiology and behavior in advance of the varied demands of the day/night cycle. Such internally generated daily rhythms are given the name circadian rhythms from the Latin *circa* (about) and *dies* (day).

Jet lag is a recognized sleep disorder that results from crossing time zones too rapidly for the circadian rhythms to keep pace. They are normally synchronized to the solar light-dark cycle and promote alertness during the day and sleep at night. Unfortunately, the clock is slow to reset, so that after time zones have been crossed, the signals for sleep and wakefulness do not match the local light-dark and social schedules. The incidence of jet lag disorder is unknown, but it presumably affects a large proportion of the more than 30 million travelers who embark from the United States each year for destinations that necessitate flights across five or more time zones. For most people, it is more difficult to travel east than west because the internal factors of the biological clock are typically longer than 24 hours and it is easier to lengthen the day than to shorten it. Tolerance to the effects of jet lag appear to decrease with increasing age. Symptoms of jet lag include insomnia and daytime sleepiness, diminished physical performance (travel fatigue), cognitive (the ability to think) impairment and gastrointestinal disturbances. Travel fatigue can be reversed within a day or two with adequate diet, rest and sleep, but other symptoms persist until the circadian rhythm is realigned.

Disturbances in circadian rhythms can also occur when workers change shifts and is associated with blindness. Night workers can have almost continuous jet lag symptoms, potentially leading to insomnia, heart disease, obesity and diabetes. Melatonin, a natural hormone that plays a role in sleep, may help these conditions, but there is a lack of long term studies to prove safety for extended use.

Chronotherapeutics or chronotherapy is a relatively recent addition to pharmacy practice based on circadian rhythms. Researchers have found that giving some drugs at specific times can improve effectiveness and reduce side effects. One study, for example, found that higher levels of anti-flu antibodies are produced when vaccinations are given in the morning. Cholesterol lowering drugs are most effective when taken at bedtime and the Food and Drug Administration recommends administering them at night.

Chronotherapy is particularly useful when considering the time for administering drugs used to treat cancer or autoimmune diseases like multiple sclerosis. Timing their regimen can significantly affect their toxicity and effectiveness. One example is the drug cisplatin. It has been found that giving the drug at the time of day when the patient's urinary output is highest reduces the effect on the kidneys. This is not surprising as cancer can disrupt the biological clock.

Cancer is not the only disease likely to be affected by the biological clock. Clinical trials have shown antihistamines are most effective when taken at night or early in the morning. Other studies established that inhaling corticosteroids at bedtime is most effective in combating allergy symptoms. Certain drugs used for high blood pressure should also be taken at night rather than in the morning. Ask your pharmacist for advice. Of course, synchronizing medications to the biological clock is easier said than done, as not everyone's rhythms are the same. One thing is certain, any clinical study for a drug should consider the time it is given.

CURIOSITY

—*The most valuable commodity!*

WHEN I HAD the opportunity to interview new job candidates I asked what, if any, books the prospective new hire had read and what his or her interests were. Reading habits are one indication of how inquisitive the individual is and a dedication to continued learning. According to the literature those individuals with strong curiosity traits are generally better and more creative problem solvers. There is a growing body of evidence suggesting that inquisitive people are more qualified to fill complex jobs and learn new skills faster. Moreover, the more curious we are about a topic, the easier it is to learn information about it. A recent article in the Harvard Business Review, entitled "From Curious to Competent" noted that curiosity, defined as a penchant for seeking new experiences, knowledge, and feedback and an openness to change is perhaps the most important of all job qualifications. While the definition appears to be complete, scientists think differently and have been investigating the effects of curiosity since the 19th century. Over time, the working definition has included "a drive state for information." The scientific definition for what constitutes curiosity is still under debate.

There is, however, new research indicating that inquisitive people provide a wide range of benefits to employers. For one, curious employees make fewer decision making errors. They are less apt to employ confirmation bias (looking for information that supports their belief rather than for evidence suggesting they are wrong) and to stereotyping people (making broad judgments). Curious people view tough situations more creatively. Studies have found that curiosity is associated with less defensive reactions to stress and less aggressive reactions to provocation. Overall, natural curiosity is associated with better job performance. Inquisitive employees make more constructive suggestions for implementing solutions to problems that occur in the workplace. It behooves companies to cultivate curiosity at all levels and treasure inquisitive minds. When triggered by design, employees think more deeply and rationally.

Curiosity may also be part of the answer to a more fulfilling life. According to a recent book by Todd Kasdan about curiosity—- it is the central ingredient. He claims that curiosity is nothing more than what we feel when struck by something novel. It draws our attention to things that are interesting and plays a critical role in the pursuit of a meaningful life. Being curious is about how we relate to our thoughts and feelings. It is not about whether we pay attention but how we pay attention to what is happening in the present. Only in the present can we be liberated to do whatever it is that we want and is a razor-thin moment when we are truly free. When we are curious we exploit these moments by being there, sensitive to what is happening regardless of how it diverges from what it looked like before and what we expect it to be in the future. There is a strong correlation between curiosity and mindfulness.

CULTIVATING CURIOSITY

Most children are naturally curious, even to the point of endangering themselves. Curiosity is unique to human beings, begins almost at birth but frequently lost as we grow older. According to Ian Leslie in his book about curiosity, the challenge is to find ways of making us continually hungry to learn, question and create. One way is to pique a person's interest in some topic. Parents are presented with the opportunity many times each day. Once curiosity is stimulated, there is increased activity in the brain circuits related to reward which enables the brain to enhance learning and retain information.

Edutopia, an educational website, suggests a number of ways to stimulate a student's inquisitiveness, (1) value and reward curiosity, (2) teach students how to ask quality questions, (3) teach skepticism by asking students to ask "why" more often, and to ask for additional evidence before accepting someone's claims as being true, (4) encourage students to tinker, (5) create opportunities for more curious and less curious students to work together in project based learning, (6) encourage students to investigate their genetic or emotional links to other cultures, and (7) most important, help parents understand the importance of curiosity in the child's development and suggest ways to foster it at home.

Final thoughts

The truly curious will be increasingly in demand. Employers are looking for people who do more than simply follow procedures competently or respond to requests. We can all learn from Albert Einstein who said that he had no special talents other than being passionately curious. Fortunately, in a free society, no one can stop us from learning. There is no limit on the pursuit of knowledge.

David Nachmansohn, Nerve Gases, and Electric Eels.

Early in World War II, the U.S. Army Chemical Corps (USACC) received alarming intelligence reports that large quantities of nerve gases were being manufactured by the Germans. Such gases, derived from organophosphates, paralyze the nervous system and are the deadliest weapons used in chemical warfare. They are odorless, stable, easily dispersed and invisible, and have rapid effects when absorbed through the skin, swallowed or inhaled. The vapor from just three drops can kill the enemy in minutes. Under optimal conditions with a brisk wind blowing at ground level, guided missiles containing the gas could destroy 90 percent of all life within many square miles. There was a desperate need for an antidote and the USACC sought help from a scientist in the United States with expert knowledge about nerve physiology. Fortunately, that individual happened to be Dr. David Nachmansohn, professor of biochemistry at Columbia University in New York City. He was given the top secret assignment to investigate the gases' action and, if possible, contrive a means to counteract the effects. Perhaps no one in history had faced a similar circumstance with so little time to respond with a solution. Even stranger is that the assignment was contingent upon the availability and the collection of electric eels.

NERVE GASES

Organophosphate compounds were first synthesized in the early 1800s when Lassaigne reacted alcohol with phosphoric acid. In the 1936, Gerhard Schrader, a German chemist, while employed by IG Fargen, discovered that several compounds, including bladan and parathion were very effective insecticides. The compounds killed insects by interrupting their nervous systems and similarly affecting the cholinergic nervous systems in humans inducing shortness of breath, choking and dimming of vision. During that same time he accidently discovered tabun—an extremely toxic compound stockpiled today as a nerve agent. During World War II, under the Nazi regime, teams led by Schrader discovered two more organophosphate nerve agents, Sarin(1938) and Soman (1944). During the war the German military banned the use of organophosphates as insecticides and instead began developing an arsenal. (Sadly, there have been recent reports of sarin being used by the Syrian government during the conflict there.)

DR. NACHMANSOHN (1899-1983)

David Nachmansohn was born in Russia, grew up in Germany, and received a medical degree from the University of Berlin in 1926. He conducted research at the Kaiser Wilhelm Institute of Biology until 1933. When Hitler assumed power, Nachmansohn fled to Paris and spent six years at the University of Paris. He came to the United States at the invitation of Yale University and after three years joined Columbia University in 1942. His greatest accomplishment was identifying isolating and analyzing chemicals responsible for the complex process by which impulses are generated along nerve and muscle fibers.

Nachmansohn was influenced by the work of Sir Henry Dale in the 1930s who proposed that acetylcholine transmits nerve impulses across junctions between neurons and muscle and that acetylcholine was rapidly hydrolyzed (inactivated) by an enzyme, acetylcholine esterase. He soon discovered that acetylcholine esterase was present in high concentrations in many different types of excitable nerve and muscle fibers. Earlier Nachmansohn had read an article describing the electrical organs of fish and managed to obtain tissues from electric eels to study. He confirmed the fact that their tissues contained exceedingly high concentrations of acetylcholine esterase and he used his findings to discover an antidote for nerve gas poisoning. It was the first compound designed on purely theoretical grounds.

ELECTRIC EELS

Electric eels are closely related to catfish, their scientific name is *Electrophorus electricus*. They are found South America, particularly in the Amazon and Orinoco River basins, and eels can attain a length of up to 8 feet and weigh up to 45 pounds. The body is elongated and cylindrical, almost without scales, with a flattened head. An eel has multiple cells (electrocytes), numbering in the hundreds of thousands, along its body that create a change in potential of up to 600 volts, with a typical current of about 1 ampere. (600 volts has the same potential as 360 D-cell batteries connected together and is 5 times that of a U.S. electrical outlet.) The electric eel is one of just a few species using electrical discharges that act like a biological taser to capture prey and defend against predators. For humans, a single jolt could incapacitate a person long enough to cause him or her to drown even in shallow water. Electric eels were useful in early understanding of electricity and helped to inspire Volta's invention of the battery in 1799. Today, efforts are also being made to build artificial cells that can replicate the electrical behavior of electric eel cells and possibly to drive future implantable medical devices. Researchers have teamed up to develop an electric eel inspired power source that would remain inside the body, completely eliminating the need for batteries. We indeed live in a surprisingly strange and shocking world.

Direct to Consumer (DTC) Drug Ads

If you hadn't noticed the amount of television time devoted to drug ads then you may not be viewing as much as I do. It appears they consume more time than the actual program. For me, the ads are becoming more and more annoying. They include therapy or preventive measures for hepatitis C, psoriasis, blood pressure, depression, cancer, urinary frequency, erectile dysfunction, influenza, shingles, Crohn's disease, Parkinsons and others. Names of the drugs are also becoming more and more bizarre and as unpronounceable as the generic identification. (The generic name of the drug must accompany the trade name.) Because the ads must include risks, the adverse effects consume most of the commercial. Some of the risks are beyond the comprehension of the average viewer. Because air time is expensive and because side effects are infinite no one can really gain all of the knowledge needed to reasonably discuss the drug with a prescriber. Some of the ads advise you to visit your gastroenterologist, as if each of us know who that might be.

Because all television viewers are exposed to direct to consumer drug advertising, they should know something about it. For example, most countries do not permit companies to advertise directly to consumers. The United States is the only country other than New Zealand that permits advertising of prescription products. Companies in the United States can also target other promotional materials directly to the consumer, such as materials distributed in a physician's waiting room. The Food and Drug Administration (FDA) requires that all consumer directed information meet the same criteria for accuracy, truthfulness, fair balance and full disclosure as materials intended for healthcare professionals. Under a policy announced by FDA in August 1997, and repeated in August 1999, advertising for prescription drugs on commercial television must make adequate provision for the consumer to obtain further information via the internet, a toll-free number or some other means. The FDA encourages disease-oriented or help-seeking advertisements to consumers that does not mention any product name, but that identifies a disease condition and urges consumers to visit their physician.

The litany of side effects that are included in television ads are extensive, some more serious than others. Side or adverse effects must be contained in drug labeling (package inserts), and include all reports obtained during clinical trials whether or not they are drug related. The side effects vary from something manageable like drowsiness or a runny nose to sudden death. Steve Martin, actor, musician, author, in his book "Pure Drivel" found the inclusion of side effects amusing and wryly added some of his own. Examples include:

- If a fungus starts to grow between your eyebrows, call the Guinness Book of Records.

- Do not operate heavy machinery while taking the drug, especially if you feel qualified for a desk job.

- If bowel movements become greater than twelve per hour, consult your doctor, or any doctor, or anyone who will speak to you.

- Under no circumstances, eat yak.

- Discontinue immediately if you feel your teeth are receiving radio broadcasts.

- Flotation devices at sea will become pointless, as the user of this drug will develop a stone-like body density; therefore, if thrown overboard, contact your doctor.

- Do not sit on pointy conical objects while taking this drug.

- While taking this drug, you might want to wear something lucky.

Drug companies that advertise directly to consumers are convinced that such promotion is effective or they would not continue the practice. Direct- to- consumer advertising is also tax deductible. However, such deductions may be on the legislative chopping block. Senator Claire McCaskill, D-Mo., crafted an amendment that removes the deduction, but it has not yet been included in the final Senate bill. Previous efforts aimed at removing the pharmaceutical ad tax exemption entirely. Two years ago, the American Medical Association called for the banishment of DTC ads, with pharmacists following that lead shortly thereafter, More recently, the National Academies of Sciences, Engineering, and Medicine in a paper addressing affordability, discussed barring DTC advertising from the list of tax-deductible business expenses. The paper also advised that manufacturers and suppliers should adopt industry codes of conduct that reduce or eliminate direct-to-consumer advertising of prescription drugs and should increasingly support efforts to enhance public awareness of disease prevention and management.

There are a number of pros and cons about drug advertising. Proponents mention free speech issues, encouragement to gain medical advice, patient compliance, revenue that can be used for research and development etc. On the other hand, such advertising can promote use before long term safety data are available, encourage over medication, weaken relationships between doctor and patient, and increase drug costs.

I look at the ads from another perspective, particularly the one where the alleged patient wonders whether he is getting the best treatment for his condition—- I thought that was the role of, and why we visit the doctor.

DOCTOR FISH

—*A cure for psoriasis?*

IN MY PREVIOUS columns I have discussed a few insects, bacteria, and animals that have a medical application. Last month I mentioned that zebrafish are being used in genetic screening and drug discovery. There is one other aquatic species to add, a small fish called *Garra rufa*, now being used in an experiment to treat individuals with skin diseases like psoriasis. Currently it is employed to perform pedicures. As such, the fish consumes dead skin on people's feet and leaves newer skin exposed—thus the name "doctor or nibble fish."[3] Fish pedicures fortunately have not been implicated in any associated illnesses but there have been a number of outbreaks of bacterial infections related to nail salon foot baths. They have resulted in wound problems, septicemia, boils and scarring. Because of these problems more than 10 states have banned the use of fish pedicures. Banning is based on at least one of the following reasons:

- The fish pedicure tubs cannot be sufficiently cleaned between customers when the fish are present.

- The fish themselves cannot be disinfected or sanitized between customers. Due to the cost of the fish, salon owners are likely to use the same fish multiple times with different customers, which increases the risk of spreading infection.

- Chinese *Chinchin,* another species of fish that is often mislabeled as *Garra rufa* and used in fish pedicures, grows teeth and can draw blood, increasing the risk of infection.

- According to the U.S. Fish and Wildlife Service *Garra rufa* could pose a threat to native plant and animal life if released into the wild because the fish is not native to the United States.

- Fish pedicures do not meet the legal definition of a pedicure.

- The fish must be starved to eat skin, which might be considered animal cruelty.

DESCRIPTION

Doctor fish is the name given species of *Garra rufa*. The species occurs in the river basins of the Northern and Central Middle East, mainly in Turkey, Syria, Iraq and Iran. Although Garra rufa are native to various Middle Eastern countries, they are largely concentrated within the river beds and hot springs in Turkey. It is thought that it was in Turkey that their unique healing properties were discovered and where they were granted the nickname.

Doctor fish spas

During the 1960's health spa's were built near the habitat of the *Garra rufa* so that visitors could benefit from the foot therapy. In 2006, doctor fish spas opened in Hakone, Japan, and in Umag, Croatia, where the fish are used to clean the bathers. There are also spas in resorts in China, Belgium, the Netherlands, South Korea, Singapore, Bosnia-Herzegonia, Hungary, Slovakia, India, Thailand, Cambodia, Indonesia, Malaysia, the Philippines, Hong Kong, the Czech republic, Spain, France and Norway. The procedure is legal in Quebec, with a few clinics in Montreal. In 2010, the first UK spa opened in Sheffield, England.

Ichthyotherapy

According to Wikipedia, ichthyotherapy can be defined as the use of fresh water or marine organisms as agents of skin wound/condition cleansing. The name comes from the Greek name for fish—ichthys.

The history of such treatment in traditional medicine is sparsely documented. There is widespread use of such fish in India, particularly in rural areas. The benefits were first observed in Kangal, Turkey - therefore it is also called the Kangal Fish.

Patients suffering from psoriasis have benefited from the doctor fish treatment, which involves lying in the ponds and letting the fish eat the scales and loose skin on the affected areas. In fact such is the popularity of the treatment for skin conditions that Kangal became a health resort. In one published study, 67 patients diagnosed with psoriasis underwent three weeks of ichthyotherapy in combination with short term ultraviolet A sunbed radiation at an outpatient treatment facility. Patients were required to stay in treatment tubs for 2 hours a day. Each patient was allocated a personal bathing tub, and the fish only came in contact with a single patient. The study, of course, was limited by the relatively small number of patients treated and by lack of a control group. Randomized trials would be needed to compare the ichthyotherapy treatment with controls and to assess treatment with standardized health-related quality of life questionnaires.

The study cited is certainly not definitive. However, in light of the wide spread distribution of psoriasis and the fact that there is no cure, makes the finding promising.

Possible mechanisms of action

Several mechanisms have been suggested regarding the observed efficacy of ichthyotherapy. One obvious mechanism is the physical contact with the fish, which feed on the desquamating skin, thus leading to a rapid reduction of the scales of psoriasis. The fish seem to prefer affected instead of healthy skin. Another suggested mechanism could be that the simultaneous removal of scales by the fish facilitated the penetration of UV rays to the dermis. This exposure of lesions may explain the better outcome of combined ichthyotherapy/UVA treatment when compared to the poor results of UVA sunbed treatment alone.

DRUGS FOR ELDERLY PATIENTS

—The Beers Criteria

ONE OF THE courses taught in pharmacy school is pharmacokinetics, a branch of pharmacology concerned with movement of drugs into, through and out of the body. The basic principles are absorption, distribution, metabolism and excretion. As originally used, these terms represented by the acronym ADME, included the drug entering the body (A), moving about the body (D), changing with the body (M), and leaving the body (E). This acronym was first used in the early 1960s and initially taught to pharmacy students in the early 1970s. Understanding these processes has changed the way drugs are administered to elderly patients (ages 65 and older). The reason is that aging involves progressive impairments in the function of multiple organs. Age changes both the function and composition of the body. The most important pharmacokinetic change in old age is a decrease in the excretory capacity of the kidney. Renal excretion is decreased up to 50 percent in about two thirds of elderly patients, and confounding factors such as high blood pressure and coronary heart disease also account for a decline in kidney function. Diminished ability by the kidneys means longer periods of drug effects and possible toxicity. Clearance by the liver of some drugs can also be reduced by up to 30 percent and drug effects are more persistent. Delayed drug excretion can result in adverse drug reactions and in elderly patients the overall incidence is two to three times that found in young adults.

The entire subject related to proper dosage is complicated by coexisting diseases, a loss in water content, an increase in fat content and multiple drug therapy. (It has been found that 36 percent of older patients take at least 5 prescription drugs.) Drug use is greatest among frail older adults, hospitalized patients and residents in nursing homes, where the typical patient takes 7 to 8 different drugs regularly. Women take more drugs than men.

The central nervous system is especially vulnerable in the elderly, drugs that affect brain function (anesthetics, opiods, anticonvulsants, psychotropics) must be used cautiously in this age group. Drug plasma levels should be monitored if possible and such technology is now available.

One of the first comprehensive articles related to managing drug therapy in the elderly was published in 1989 in the New England Journal of Medicine. The article included a number of guidelines for prescribing effectively for that age group. According to the authors, increasing the knowledge of the action of drugs in the elderly and improving communication between patient and physician will greatly improve the overall care of the older patient. Fortunately, since then new consensus criteria as well as clinical practice guidelines have been developed with the potential to improve the quality of care for the aged. The criteria for determining inappropriate drug use began with publication of a seminal article in the Nursing Home Medication Use journal in 1991. The lead author was Dr. Mark Beers and the subsequent criteria bear his name. The article was followed by widespread efforts to educate clinicians about the criteria and to use them in quality improvement activities that have meaningful impacts on the quality of care of older adults.

The American Geriatrics Society continues to publish what is now described as "A Guide for Patients, Clinicians, Health Systems, and Payers", the principles are the basis for the "AGS Beers Criteria." It contains a list of criteria providing recommendations for medications that should often be avoided for older adults. The criteria contain guiding principles for patients with multiple chronic conditions that summarize the pitfalls when treating complex, frail older adults. Principles include the proviso that (1) medications listed in the Beers list are potentially inappropriate rather than not definitively inappropriate, (2) that medications listed in the criteria are not intended for use among patients at the end of life because unique prescribing considerations must be used at that time, (3) medications included in the criteria must be adjusted based on benefit to risk calculations, (4) inappropriate medications are identified together with safer alternatives, (5) the Beers Criteria should be a starting point for a comprehensive process of identifying and improving medication appropriateness and safety, (6) the Beers criteria should not be excessively restrictive, and last, that the Beers Criteria should not be equally applicable in all countries. The reason is that certain medications are unavailable worldwide. It should be noted that the Beers Criteria is a list of potentially inappropriate medications, and the inclusion of a drug on that list does not mean that it should never be used in the elderly. The list is meant to be a guideline for identifying medications for which the risks may outweigh the benefits in older patients. The prescriber's clinical judgment of each individual patient's needs should always be considered. Care givers trusted with care of the elderly may wish to discuss the Beers Criteria with their health care providers.

Dying

*—Ssomething to think
and not worry about.*

It's not that I am afraid to die. I just don't want to be there when it happens. -Woody Allen

A RECENT ISSUE of The Economist contained some startling information on the aging population. For example, the average 65 year old in the rich or developed world (as opposed to the third world) can now expect to live another 20 years, half of them free from disability. However, from the age of 80, again in the developed world, one person in five will be afflicted with some form of dementia, one in four will suffer from vision loss and four in five will develop hearing problems. Of those who make it until 90, the majority will have at least one health problem that counts as a disability; many will have multiple ones. In America today, a 70 year old man has a 2 percent chance of dying within a year; in 1940 this milestone was passed at age 56. In 1950, just 5 percent of the world's population was over 65; in 2015 the share was 8 percent, but by 2050 it is expected to rise to 16 percent. Of note is that in Britain there were just 24 centenarians in 1917, today there are nearly 15,000. (A centenarian is 100 years of age or older.) It is estimated that half of all children born in the developed world during this century will live to at least 100. One new study finds no evidence that the proposed maximum lifespan of around 115 years has stopped increasing. To the contrary the study demonstrates that any such maximum has yet to be identified. But no matter what aging does to us we all have to die and death is a topic most of us wish to avoid talking about. Because It is inevitable, the more we learn about dying the less painful it might be.

HOW WE DIE

One of the best books on the subject of dying, written by Sherwin B. Nuland, demythologized the process by helping rid ourselves of the fear of the unknown. According to Dr. Nuland everyone wants to know the details of dying, though few are willing to say so. Moreover, modern dying frequently takes place in a hospital where it can be hidden, cleansed of its organic blight, and finally packaged for modern burial.

Dying begins at birth, of course, but biological processes that throughout life have been making replacement parts for dying structures within each cell can no longer do their job. After a life time of regenerating spare parts, the nerve and muscle cells capacity for regeneration gradually shuts down. The rapidity of circulation diminishes. Cardiac cells cease to live and the heart loses strength. Each heart beat pushes out less blood than it did a year earlier. As the pump ages, its inner lining and valves thicken. Not only the heart itself but the blood vessels are affected by the passing years. Before long every organ is getting less nourishment that it needs. The normal kidney loses some 20 percent of its weight and develops scarring. Thickening of the tiny blood vessels inside the kidney decreases blood flow and the kidney loses its ability to remove excess sodium or to retain it in the body when needed. The bladder is also affected. As it ages the bladder loses its distensibility and can no longer hold as much urine as before. Old people need to urinate more frequently. Like muscles of the heart, brain cells are unable to reproduce. For every decade after 50, the brain loses 2 percent of its weight. Even the immune system is not immune to aging.

In essence, aging results from an accumulation of unrepaired cellular and molecular damage and the limitations in cell maintenance and repair functions. In particular, those of DNA and proteins. The maintenance of DNA integrity is a challenge to every cell for damage leads to the absence of key proteins, and to the synthesis of proteins in the wrong cells at the wrong time. Such damage accumulates throughout life from the earliest time when body cells and tissues begin to form. The same is true for mitochondria (the energy factor in cells).

The Symptoms of Dying

It is comforting to know that symptoms related to dying such as the death rattle (caused by saliva collecting in the throat), air hunger, and agitation are generally not uncomfortable to the dying person. All are well treated by medications and with hospice—it is rare to die in pain.

FINAL THOUGHTS

The great sixteenth century essayist Michel de Montaigne was once on the verge of dying after an accident and found himself gasping for air, and attempting to pound on his chest to breathe. Fortunately he recovered. He later reflected that despite the trauma, he began to grow languid while feeling like he was being carried aloft on a magic carpet. From this he found that learning to die is not necessary. He noted, "If you don't know how to die, don't worry; nature will tell you what to do on the spot, fully and adequately. She will do the job perfectly for you; don't bother your head about it."

ELEPHANTS

—Large, leathery, lumbering, lovable animals.

ELEPHANTS INDEED ARE large, leathery, lumbering, and loveable, a reminder of what prehistoric animals were like. But there is more to elephants than just alliteration. For one, elephants may be a lot smarter than we are. Second, their trunks have a number of exceptional properties. Third, because elephants can weigh as much as eight tons, they should be particularly prone to cancer, but they are not. And unlike a number of animals, elephants are attentive, social, and generally non-aggressive.

INTELLIGENCE

An elephant's brain is the most sizeable of all, it weighs just over 5 kg or 11 pounds. The human heart weighs about 3 pounds. Although the largest whale is 20 times the size of an elephant, its brain is just under twice the size. The need for such a large and complex organ becomes clear when considering their behavior and ability. Elephants are capable of a range of emotions, including joy, playfulness, grief and mourning. In addition, elephants can learn new facts, mimic sounds they hear, self medicate, play with a sense of humor, perform artistic activities (paint), use tools and display compassion and self awareness. Part of the reason is the structure of an elephant's brain. The neocortex is highly convoluted, as it is in humans, apes and some dolphins. This is generally accepted to be an indication of complex intelligence. The elephant is one of the few creatures (together with humans) that is not born with survival instincts, but needs to learn them during infancy and adolescence. Elephants and humans have a similar lifespan, and plenty of time, approximately ten years to learn before they are considered independent adults. Their behavior indicates that elephants can also identify language, if the voice belongs to a person who is likely to pose a threat, an elephant will switch into a defensive mode.

THE TRUNK

According to an article in the New York Times, an elephant's nose is the most unusual feature of them all. It is the oddest of all dangling appendages, actually a supernaturally strong, skin covered slinky that has fine motor skills, sensitivity, and caressability (for reassuring or comforting). Elephants have more scent receptors than any other mammal, they can help soldiers avoid minefields in light of their ability to detect TNT. So sensitive is an elephant's trunk that is more capable than a bloodhound's nose and able to smell water from several miles away. The nose or trunk is both and upper lip and a nose, with two nostrils running through the whole thing. At the trunk's tip, African elephants have two fingers while Asian elephants have one. The dexterity of the fingers allows an elephant the ability to deftly pick up a single blade of grass or hold a paint brush. An elephant's trunk has eight major muscles on either side and 150,000 muscle bundles in all. It is so strong that it can push down trees and lift up to 300 kg (660 pounds). The trunk can stretch, and reach branches 20 feet high, and by extending out of the water like a snorkel, it enables an elephant to cross bodies of water too deep for other less equipped animals. As a water tool, the trunk can suck up to 10 gallons of water a minute and hold up to two gallons at a time. While not related to the trunk, the elephant's large size ears help to flap the heat away.

Cancer

It takes billions, perhaps trillions of cells, to compose an elephant. All of those cells arise from a single fertilized egg, and each time a cell divides, there is a chance for a mutation to occur, one that may lead to a tumor. Strangely, elephants are not more prone to cancer than smaller animals and research suggests that they get less cancer than humans. Recently researchers reported that elephants protect themselves with a unique gene that aggressively kills off mutant or DNA damaged cells. Elephants have evolved unusual p53 anticancer genes, the genes that makes a protein that senses damage. While humans have one copy of the gene, elephants have twenty copies. In elephants the p53 proteins switch to another gene called LIF6, it is even more potent in detecting damaged cells than the p53 variety. No other animal has that gene. Somewhere in the course of elephant evolution, a cellular mutation inserted a genetic switch to LIF6, enabling the gene to be activated by p53.

FINAL THOUGHTS

Unfortunately, due to poaching (ivory tusks) and habitat destruction, the elephant population is rapidly declining. The tusks (overgrown versions of the upper lateral incisors) designed for defense may be the reason for the elephant's eventual demise. In Asia, it is estimated that less than 50,000 elephants remain. Elephants are also a step further from extinction in Africa. In the early 1800s, it was estimated there may have been 26 million elephants in Africa alone, those numbers today are a tiny fraction of that population.

Emery Andrew Rovenstine MD

—*Someone to be proud of.*

Two days before my column about anesthesia was published in this paper I attended an anniversary party for two friends here in Warsaw. When I described the contents of the article and why I had a long term interest in drugs used in surgery, one of the other guests modestly mentioned that his uncle, Emery Rovenstine, worked in the anesthesiology department at a New York City hospital. His comment turned out to be a remarkable understatement about someone who had been the world's foremost anesthesiologist. Among his many honors, Dr. Rovenstine was chair of the Department of Anesthesiology, New York University Medical Center and Director of Anesthesia, Bellevue Hospital, a founder and past president of the American Board of Anesthesiology, past president of the American Society of Anesthesiologists, and recipient of the Society's 1957 Distinguished Service Award. He was a dominant figure in anesthesiology in the United States between 1935 and 1960. His extraordinary research and writings had an enormous impact on the specialty and continue to this very day. Dr. Rovenstine's accomplishments were not limited to his work in the United States, he accepted a guest professorship at Oxford University in England, and later at the University of Rosario in Argentina. According to Wikipedia, he also received visiting appointments in Bohemia, Canada, Cuba, Czechoslovakia, France, Japan, Mexico and South Africa, and he was inducted into the medical society of each respective nation.

Dr. Rovenstine was born in Atwood, Indiana, in 1895, where he clerked in his father's grocery store. He briefly attended Winona college in Winona Lake, and taught high school before moving to Wabash College, where he graduated in 1917. Upon graduation, Rovenstine enlisted in the U.S. Army and served in France during World War I. During the three years active duty he spent much of his time in charge of an engineering demolition squad where he was an eye witness to battlefield pain and suffering. After returning home, he spent several years teaching and successfully coaching basketball at LaPorte High School.

Rovenstine thought of his war time experiences and decided to attend the University of Indiana medical school. He received an MD degree in 1928. In 1930, after struggling to maintain a general practice in LaPorte, he took a faculty position at the University of Wisconsin—Madison, where he served as an assistant professor of anesthesiology. He worked with his mentor, Dr. Ralph Waters, to develop the use of cyclopropane as an anesthetic agent and they were the first to use it on human subjects. (Cyclopropane is an alternative to using ether or chloroform.) Dr. Waters was one of the country's first few anesthetic specialists, and he developed a department distinguished by its close collaboration between practice and pharmacology and physiology. By the 1930s, the university was recognized as the "Mecca of Anesthetics."

In light of his accomplishments, in 1935 Rovenstine was appointed chair of the department of anesthesiology at Bellevue Hospital, New York City's oldest hospital. While there he was influential in shaping the department's mission and mentoring future generations of anesthesiologists. One example was his work in developing the technique for using muscle relaxants like curare during anesthesia. The addition of muscle relaxants to the regimen has changed the practice of anesthesia and increased the range of surgical procedures. Today muscle relaxation is an irreplaceable part of anesthesia. He was also a dominant figure contributing to regional anesthesia—using local anesthetics to block sensations of pain for a large area of the body, such as an arm or leg or the abdomen. Regional anesthesia (nerve blocks) allows a procedure to be done on a region of the body without the patient losing consciousness. Two of the most frequently used are spinal anesthesia and epidural anesthesia, which are produced by injections made in the appropriate areas of the back. Other examples where regional anesthesia is used include thoracic surgery, ophthalmology, gynecology and for surgery of the stomach, intestines or liver.

Dr. Rovenstine also did seminal and important research on a means of disposing of, or preventing carbon dioxide from being re-breathed when inhalers are used for inhalational anesthesia. This is a critical factor to ensure that anesthetics are used safely and effectively.

During World War II, Rovenstine served on the Army Advisory Board and was responsible for an order to Army general hospitals placing anesthesiologists in charge of operating rooms.

Dr. Rovenstine, was indeed, an important pioneer in advancing surgery and surgical technique, a gift we recipients should be thankful for. He died at age 65, at the height of his career, the world's most respected anesthesiologist. Most assuredly, a local hero.

Escherichia coli

—Friend or foe?

SEVERAL YEARS AGO a deadly outbreak of *Escherichia coli* infection in Europe was linked to contaminated bean and seed sprouts from an organic farm in Germany. There were 42 deaths and approximately 3900 individuals infected with the rare and super toxic 0104:H4 strain of the bacteria. More than 780 persons developed kidney failure. E. coli infections are also quite common in the United States and elsewhere. According to the Centers for Disease Control and Prevention (CDC), E. coli has also been associated with food poisoning from consuming bologna, cheese, hazel nuts, romaine lettuce, poultry, beef, pizza, and cookie dough. E. coli is also the most common etiologic gram negative organism responsible for U. S. hospital acquired urinary tract infections.

The fact that E. coli bacteria can come in many toxic strains or serotypes which reproduce at astronomical rates is the basis for their potential danger. The organism can double its population in less than two hours under the right conditions which means that it has the potential to make people, especially children and the elderly, very sick. The explosive population rate is also one of the reasons E. coli can be used for genetic research. All

E. coli strains share the same underlying biology, but they range from being harmless and beneficial to being extremely dangerous pathogens. The well known strain K-12, for example, is harmless. Other strains are a different story and books have been written that describe their mechanisms of virulence.

E. coli was first described in 1885 by Theodor Escherich, a German pediatrician, in a monograph on the relationship of intestinal bacteria to the physiology of digestion in the infant. (The organism was isolated from the diapers of healthy babies.) He called it "bacterium coli commune." At that time, E. coli was merely one of a rapidly growing list of species of bacteria that scientists were discovering. In 1919, the name *Escherichia coli* was proposed in his honor but it was not officially recognized until 1958.

THE ORGANISM

Escherichia coli is a typical member of the Enterobacteriaceae family that have their principle habitat in the bowel of humans and animals. It is a short, straight Gram negative bacteria. In nature it is found in soil, water or at any other site it can reach from its primary habitat, usually by fecal contamination.

E. coli is a fairly typical bacterium about 1 micron wide and 2 microns long. Thus a billion of them can be packed into a volume of a few cubic centimeters. They can be frozen alive and in the frozen state they can persist almost indefinitely without any serious loss. At a very low temperature, such as in space, many of them would likely survive for well over ten thousand years. The characteristics of this organism make E. coli an ideal laboratory research tool.

As mentioned above, E. coli is widely disseminated throughout the food chain. In the 1990s and into the early 21st century, the majority of food borne E. coli outbreaks were caused by the consumption of contaminated ground beef. Numerous outbreaks and massive recalls of contaminated meat products continue to plague the meat industry and the public. Water intended for recreation and for human consumption can also become contaminated. Other means of transmission include person-to-person and animal-to-person contact.

In the early twentieth century, scientists began to study harmless strains of E. coli to understand the nature of life. Now more is known about E. coli than about any other organism in the biosphere, including humans, and the genome for E. coli is one of the most extensively mapped of any organism. Generations of researchers have probed into the existence of the organism, carefully studying most of its 4000 –odd genes and discovering more and more about evolution. Through this work scientists can see an ancient history we share, a history that includes the complex features in cells, the common ancestor of all living things, a world before DNA. With the knowledge gained from E. coli, genetic engineers now transform corn, pigs and even fish. E. coli has also been used to define the molecular and cellular mechanisms underlying how microbes cause disease.

Carl Zimmer, an award winning science writer, perhaps

best describes the friend and foe aspects for E. coli in the following paragraph :

"E. coli may seem like an odd choice as a guide to life if the only place you've heard about it is in the news reports of food poisoning. There are certainly some deadly strains in its ranks. But most E. coli are harmless. Billions of them live peacefully in my intestines, billions more in yours, and many others in just about every warm blooded animal on Earth. All told, there are around 100 billion billion on Earth. They live in rivers and lakes, forests and backyards. And they also live in thousands of laboratories, nurtured in yeasty flasks and smeared across petri dishes."

Falls in the Elderly

Both my mother and mother-in-law were almost 100 years of age, when they fell and suffered hip fractures. Neither event was unexpected as fall rates are extremely high for this age group. Fracture of the hip is one of the most common and medically devastating conditions affecting older persons, threatening survival as well as independence. More than 90 per cent of hip fractures occur as a result of falls, with most of these fractures occurring in persons 75 years of age or over. According to the literature, one third of the population greater than 65 years of age living at home report falling one or more times a year and one out of every 200 home falls produces a hip fracture. At ages greater than 85 and above, one hip fracture occurs for every ten falls. While the hip is the bone most frequently sited, fractures can also occur in the pelvis, vertebrae, humerus, hand, forearm, tibia and ankle.

Among persons aged 65 years or older, falls are the leading cause of death from injury. The mortality rate for falls increases dramatically with age in both sexes and in all racial and ethnic groups, with falls accounting for 70 per cent of accidental deaths in persons 75 years and older. Every 19 minutes in this country, an older person dies from a fall. In the United States, 60 per cent of fatal falls take place at home, 30 per cent in public places, and

10 per cent in health care institutions. Home falls account for the greatest mortality because the elderly stay in this environment most of the time and because there is possibility of falling from greater heights i.e., stairs. Falls are all too frequently the cause of not only death, but of long term disability as well.

Causes

There are multiple factors that contribute to hip or other fractures. The relevant issues include: (1) whether or not a fall occurs; (2) the mechanism of the fall; and (3) how well bones and soft tissues can absorb the energy of an impact once a fall is initiated. Direct causes can be intrinsic or extrinsic. Intrinsic causes relate to the patient's susceptibility including age-related physiologic changes, diseases and medications. Extrinsic causes relate to environmental hazards. Age related changes that impair postural stability and blood pressure homeostasis make older persons more susceptible to falls and fainting. In addition to age- related changes, disease related factors such as rheumatoid arthritis, cancer, Alzheimer's, Parkinson's and cardiovascular and neurologic disease further increase the risk of falls. Osteoporosis is often included as a risk factor but its role as the cause of hip fractures remains unclear. Reduced sensory acuity in sight, hearing, and smell play a role, as do altered balance and gait, delayed reaction time, and declining muscular endurance. As a result of the development of disease, the elderly often adopt a sedentary life style, become out of shape, and have difficulty adapting to environmental stress.

The use of medications or improperly used assistive devices may further increase the risk of falling. Prescribed medications, so common to elderly patients, are frequently associated with an increased risk of falls and fall related injuries. Medications that have been implicated include diuretics, cardiac drugs, corticosteroids, hypoglycermic agents, antihypertensives, psychotropics, antiparkinsonism drugs, narcotic analgesics, anticonvulsants and antihistamines. Psychotropic drugs include antidepressants, anxiolytics (especially long-acting benzodiazepines), sedative-hypnotics and antipsychotics. Their use can result in excessive sedation, psychomotor impairment, confusion, dizziness and ataxia—- side effects that can result in falls. Topical eye medications used to treat glaucoma may also increase the risk of falling. Other factors to consider include taking four or more prescription medications, or the initiation of a new drug treatment in the previous two weeks.

The extrinsic factors mentioned earlier include poor lighting, unsafe stairways, irregular floor surfaces, clutter, poorly designed furniture, etc. Frail elderly persons tend to fall and injure themselves in the home during the course of routine activities. Fracture injuries linked to walking leashed dogs have also increased significantly in the past several years.

PREVENTION

There are a number of critical steps that can be taken to reduce the risk of falls in the elderly. Included were the need to eliminate environmental hazards, improve home supports, modify medications, provide balance training, supply opportunities for socialization and encouragement and to involve the family to a greater extent in supportive measures. The National Center for Injury Prevention and Control provides a web site describing a number of factors related to falls and hip fractures among older adults. Included are a number of bullet points older adults can consider to reduce their risk of falling. They include maintaining a regular exercise program, taking steps to make living areas safer, asking doctors to review all medicines to reduce side effects and interactions and having vision checks each year. Tai Chi is an excellent, low impact way to improve balance. Also, practice standing on one foot when you brush your teeth or wash dishes.

CLOSTRIDIUM DIFFICILE COLITIS

—A very unusual treatment method.

COULD ANYONE IMAGINE a medical condition that would warrant a patient receiving a stool specimen from a donor to help treat a disease? Such treatment would seemingly belie the "Above all, do no harm" axiom thought to be part of medicine's Hippocratic oath. It just so happens that such an event occurred more than 50 years ago, continues today, and much more often. In 1958, doctors in Denver, Colorado, administered donor feces by enema to patients with colitis. The goal of infusing donor feces was to re-establish the balance of nature within the intestinal flora to correct the disruption caused by antibiotic treatment. The doctors reported immediate and dramatic responses and concluded that this simple yet rationale method should be given more extensive clinical evaluation.

It took fifty years or so to establish an association between *Clostridium difficile* infection (CDI) and pseudomembranous enterocolitis and to identify effective antimicrobial treatment methods. Despite these advances, *C. difficile* became the most commonly identified cause of hospital acquired infectious diarrhea in the

United States. The success of fecal transplants in prevention supports the concept that there are microorganisms in feces that reestablish protection.

Clostridium difficile, the agent that causes pseudomembranous colitis associated with antibiotic therapy, has been identified in recent years as a common nosocomial (hospital acquired) pathogen. It was first described in 1935. This gram positive anaerobic bacillus was named "the difficult clostridium" because it resisted early attempts at isolation and grew very slowly in culture. Although the organism released potent toxins in broth culture, the fact that it was found in stool specimens from healthy neonates led to its classification as a commensal. (A commensal is an organism that benefits from a symbiotic (associated biologically) relationship with another organism without harming it.) The pathogenicity of *C. difficile* is rooted in the fact that it is a spore forming toxigenic organism. The spore form is resistant to gastric acid and can therefore readily pass through the stomach to the intestine, where it changes to a vegetative life cycle.

Pseudomembranous colitis was first described in 1893; while the role of *C. difficile* as the cause of diarrhea was first reported in 1978. Today, the organism is the leading cause of antibiotic-associated diarrhea and pseudomembranous colitis. During the past decade, there has been an alarming increase in the incidence and severity of this disorder, with associated increases in mortality and economic cost. The most common symptoms include watery or bloody diarrhea, fever, loss of appetite, nausea, stomach pain, cramping, and tenderness. The disease strikes 500,00 Americans each year and kills 30,000. It has been reported that 96 per cent of patients with symptomatic infection had received antimicrobials within 14 days before the onset of diarrhea and that all had received an antimicrobial within the previous 3 months.

THE TREATMENT

There are a number of alternative treatment methods. Metronidazole is the drug of choice for mild to moderate CDI, while vancomycin is used for a severe episode. Both drugs have a long clinical history. Various regimens have been used and the drugs may be used from 2 to 8 weeks. The purpose of the therapy is to keep *C. difficile* vegetative forms in check while allowing restoration of a normal flora. Methods to control infection in the hospital (gloves, gowns, hand hygiene) and environmental cleaning and disinfection are equally important. Cleaning and disinfection are critical as the bacteria is sticky similar to anthrax. *Clostridium difficile* spores have an exosporium that confers a particulate adherence—chains of protein-containing substances that adhere to hands.

FINAL THOUGHTS

Fecal implants are certainly medical oddities but preliminary results have confirmed their effectiveness and are warranted in light of the dire consequences of CDI. According to a recent editorial in the New England Journal of Medicine, a recent report of successful results should encourage and facilitate the design of similar trials for other indications, such as inflammatory bowel disease, irritable bowel syndrome, prevention of colorectal carcinoma, and other disorders. As such, fecal implants herald a broad and exciting new branch of human therapeutics. Before such efforts begin in earnest, the FDA must decide how fecal implants are regulated. According to the New York Times, the heart of the controversy is a question of classification: are fecal microbiota that cure C. *difficile* a drug, or are they more like organs, tissues and blood products that are transferred from the healthy to the sick? The answer will determine how the Food and Drug Administration regulates the procedure, how much it costs and who gets the profit.

More smart men

Last year I wrote a column about two of the smartest men who lived in the modern era (1700 to the present). It was a difficult choice considering all of the brilliant people to choose from. After careful deliberation, I chose Leonhard Euler and Thomas Young for a number of reasons. Both gentlemen were experts in multiple fields, and their works and concepts are an integral part of modern engineering, astronomy, medicine, physics, and other fields as well. Euler derived modern mathematical principles and contributed to the fields of geometry, trigonometry and calculus. In calculus alone, he provided hundreds of discoveries and proofs along with many computations to simplify and clarify differential calculus, infinite series and integral techniques. He was a revolutionary thinker in such diverse fields as astronomy, acoustics, hydrodynamics, mechanics, music, ballistics, navigation and topology. His productivity was equally amazing, during his career he wrote more than 850 publications, including 18 books. Euler lost sight in one eye in his early 30s, and was nearly blind by age 60. Despite that, he continued his illustrious career and published more than 400 more articles and a major three volume work on lunar motion.

Thomas Young was a polymath, a linguist, physician and physicist who established the theory of light, color perception, anatomy, the significance of energy, elasticity and because of his uncanny knowledge of language, the study of Egyptology and hieroglyphics. The latter was instrumental in his ability to help translate the previously undecipherable and mysterious Rosetta stone. Young also worked on liquid molecule size and surface tension measurement. While he was still a medical student he discovered how the lens of the eye changes shape to focus on objects at different distances which led to the discovery of the cause of astigmatism. When asked in later years to contribute to a new edition of the *Encyclopedia Brittanica,* Young offered to write on: the alphabet, annuities, capillary action, cohesion, color, dew, Egypt, focus, friction, haloes, hieroglyphics, hydraulics, motion, resistance, ships, sound, strength, tides, waves and anything about medicine.

On further deliberation and careful thought I nominate two others to consider—Hermann Helmholtz and Joseph Leidy.

HERMANN HELMHOLTZ (1821-1894)

According to David Cahan's book, Helmholtz's scientific achievements and his philosophical reflections on science ultimately became a collection of seven thick volumes: three of collected scientific papers (containing about 175 original papers plus five or six dozen reprinted versions or translations), a three-part tome on physiological optics, a volume on physiological acoustics and music, and two volumes of essays on popular science and philosophy, as well as a six-volume (in seven) set of lectures on theoretical physics that was assembled posthumously (and apparently considerably recomposed) by several of his last students. As this publishing record suggests, Helmholtz was a workhorse and, at times, a workaholic as well. He also helped construct and single-handedly directed three scientific institutes (one for physiology in Heidelberg and two for physics in Berlin). Helmholtz helped set the direction in a range of scientific fields during the second half of the nineteenth century. He remained an inspirational and instructive figure for many twentieth- and twenty-first-century scientists.

Joseph Leidy

Joseph Leidy who has been described as the last man who know everything by author Leonard Warren. Leidy was for many years America's foremost microscopist and human anatomist, the discoverer of Trachina larva in pigs (a notable achievement at that time) and the founder of American parasitology. Actually, he introduced the idea of parasitism here. He was also the founder of American vertebrate paleontology, the first to describe the dinosaur in America and give it its present day configuration. Moreover, he discovered and identified several varieties of amoebas, plants, worms, reptilian and mammalian fossils, and was the first to transplant human cancer cells in an animal. Leidy was appointed professor of anatomy at the University of Pennsylvania, appointed surgeon at Satterlee General Hospital, the founding member of National Academy of Science, and professor of natural history at Swarthmore college. He received a number of prestigious awards and was a prolific writer and lecturer. At the University of Pennsylvania he founded the Department of Biology and was named director and professor of zoology and comparative anatomy.

Helmholtz and Leidy's intellectual gifts differed from those of Darwin or Einstein, for example, who concentrated their efforts almost exclusively on elucidating foundational theories in one field, such as biology or physics. Helmholtz and Leidy's achievements, by contrast, ranged across the physical and life sciences (including medicine), and each did transformative work in others. Few, if any, individuals have accomplished as much in their lifetimes. It should be noted that there was no true analytical measurement method to determine intelligence in the 19th century, and even if there were IQ tests, they would not be the only measure of genius. According to historian Roy Porter, "there seems to be no common denominator except uncommonness."

Frances Oldham Kelsey and Helen Brooke Taussig

—*Two memorable advocates for drug safety*

Dr. Frances Kathleen Oldham Kelsey worked at the Food and Administration for years, retired at age 90 in 2005, and died in London, Ontario, on August 7, 2015, at age 101. Dr. Kelsey was a pharmacologist from the agency who helped save the United States from the thalidomide disaster, but also prompted widespread and justifiable concern about the inadequacy of drug regulations in the 1960s. Dr. Kelsey's efforts in premarket review are part of the fascinating history of food and drug law and the effect individuals can have in changing it.

Dr. Taussig

Not as well known perhaps was the work done by Dr. Helen Taussig. When in 1962, she heard about the tragedy unfolding in Europe due to thalidomide, she traveled there to examine affected babies and learn about their medical history. Having confirmed that thalidomide was the cause, she returned to the United States, when she testified, lectured and wrote in opposition to the pharmaceutical industry until the FDA refused to approve the drug. Without the indefatigable work done by Drs. Kelsey and Taussig, thalidomide would have been approved here in the early 1960s.

Dr. Taussig was one of the few women to apply and graduate from Johns Hopkins University Medical School, and she obtained her degree there in 1927. Her work in pediatric cardiac surgery was instrumental in her appointment as chief of the department, a position she held until her retirement in 1963. Many of the procedures she helped develop are still in use today. She died in 1986.

A LITTLE ABOUT THALIDOMIDE

Thalidomide was synthesized in West Germany in 1953 and, because it had no pharmacological effect in laboratory animals, the original company working with the drug discarded it. Subsequently, the West German firm Chemie Grunenthal undertook development of the compound, but once again thalidomide showed no effects in animals and appeared to be nontoxic. (It seemed impossible to find a dose high enough to kill a rat.) Because the structure of the molecule suggested it could be effective as a sedative acting promptly to provide a deep, all night sleep without a hangover. Given the trade name Contergan in 1960, it became the most commonly used sleeping remedy in West Germany, available without a prescription. It was thought to be as safe for humans as for animals.

By the early 1960s, pharmaceutical companies in countries other than West Germany began to market the drug.

At the height of its demand, thalidomide was being produced by 14 firms under 15 different trade names.

On September 8, 1960, the Canadian branch of William S. Merrell Company of Cincinnati, Ohio applied to the Food and Drug Administration for approval to sell their version of thalidomide, Kevadon, in the United States. The review was passed on to Dr. Kelsey, a new medical officer. Because the drug had already been sold to pregnant women in Europe for morning sickness, and the application seemed to be routine, it appeared to be readily approvable. Existing laws at that time held that after an application had been submitted, the agency had sixty days in which to decide that the drug was safe for the proposed use; if the FDA did not respond, the drug was automatically approved. But some of the data on drug safety troubled Dr. Kelsey and she asked Merrell management for more information.

The application for thalidomide landed on her desk on September 12, 1960, one week after she reported to work. She was concerned that no lethal dose for rats could be found as it suggested that the rats were simply not absorbing the drug. Then there was a question as to why thalidomide was an hypnotic in humans and not able to induce sleep in animals. Dr. Kelsey waited until the 58th day after her review before declaring the application incomplete, and thus ineligible for submission. This meant that a new application had to be filed which provided more time for Dr. Kelsey to ponder and reject it.

Congress, in light of the thalidomide tragedy reexamined the new drug approval process and promulgated the

Drug Amendments of 1962. The amendments were the most comprehensive revision of the drug regulatory law since passage of the 1938 Act.

Final thoughts

Thalidomide is also newsworthy and is currently being used for a number of indications with a host of safeguards. It was initially approved on July 16, 1998 to treat a clinical form of leprosy and later for multiple myeloma.

Francis Galton

—Another polymath!

I RECENTLY ATTENDED a lecture where the speaker discussed investment plans, hedge funds and the stock market. Most of his talk was beyond my comprehension and I have no plans to be wealthy. I did learn one thing, however, which stemmed from a demonstration he gave. The speaker showed everyone a large jar full of M & Ms and asked all of the attendees to guess how many there were. The answers ranged from 3000 to 30,000, and the correct number was somewhat in the middle. According to the speaker, Sir Francis Galton was the first person to predict that the correct answer is always the average of guesses despite how far apart the highest guess is to the lowest. I wasn't sure if the speaker was correct, but did have some recollection about Francis Galton from my general reading on statistics and decided to learn more about him. Here is what I found. I learned that in 1907, Galton asked 787 English villagers to guess the weight of an ox. None of them guessed the right answer, but when Galton averaged their guesses, he arrived at a near perfect estimate. This was the classic demonstration of the "wisdom of the crowds", where groups of people pool their abilities to show collective intelligence. The speaker at the meeting was right.

BIOGRAPHY

Galton was born on February 16, 1822, in Birmingham, England, and was the half cousin of the famous naturalist Charles Darwin. Galton and Darwin shared the common grandfather Erasmus Darwin, the famous naturalist and philosopher He originally intended to become a physician studying at King's College London, Trinity College and the University of Cambridge, but upon the death of his father, Galton inherited a fortune that allowed him to leave his medical studies and travel. His expeditions through unexplored parts of Africa won him a medal from the French Geographical Society and election to the Royal Society.

ACCOMPLISHMENTS

Galton was a British science writer and an amateur researcher of the late nineteenth century who contributed to the fields of statistics, experimental psychology, and biometry. He has been called a statistician, polymath, sociologist, anthropologist, eugenicist, tropical explorer, geographer, inventor, meteorologist, proto-geneticist, and psychometrician (someone who practices the science of psychological measurement). Not withstanding his questionable theories of racial differences, he is considered one of the world's most productive men owing to the breadth of his work.

Galton is widely regarded as the originator of the early twentieth century eugenics movement. Eugenics is a science that deals with the improvement (as by human mating) of hereditary qualities of a race or breed. He is one of the first scientists to apply statistics to heredity and held statistics in high regard. According to Galton, "Some people hate the very name statistics, but I find them full of beauty and interest. Whenever they are not brutalized, but delicately handled by the higher methods, and are warily interpreted, their power of dealing with complicated phenomena is extraordinary. They are the only tools by which an opening can be cut through the formidable thicket of difficulties that bare the path of those who pursue the Science of man."

Galton was the inventor of scientific meteorology and developed the first weather map, publishing Meteorgraphica, or methods of mapping the weather in 1863. In the 1890's Galton established that fingerprints did not change as a person ages and confirmed that they could be used as a unique method to identify a particular individual. He also identified eight types of fingerprint patterns and a classification system still used to day. The results of his work on classification and indexing was summarized in a 200 page book entitled simply *Finger Prints*.

To further his versatility, in his spare time he invented the Galton Whistle to evaluate hearing ability, by determining that the normal upper limit of human hearing was around 18 kHz. He also established that the ability to hear higher frequencies declined with age. Galton adapted his whistle to test the hearing of various animals, and it is commonly used as a dog whistle.

While in his 80s, Galton was asked to consider writing his autobiography, and he readily agreed. The result entitled *Memories of My Life,* was one his most successful books, with good reviews and a first edition that sold out within a month. The book is not a self revealing document, it is packed with reminiscences and anecdotes of the many eminent men he had the good fortune to know.

Unfortunately, according to a recent article in Nation magazine, eugenics as described by Galton, are merely the weird step-uncle of modern, scientifically grounded genetics. Galton's theories on racial inferiority must be evaluated for evidence of racism and prejudice before implementing them and harming the innocent.

Fruit Flies

—Pioneers in genetic research!

Most everyone knows that fruit flies gravitate toward bananas and most overripe fruit and almost magically appear in kitchens around the world. Less well known is that this insect species has been used by thousands of scientists around the world for more than 100 years to study genetics and developmental biology. The reasons are that fruit flies are easily cultured in the laboratory, are low cost, have a short life cycle, a simple genome (all the inheritable traits of an organism), and produce many offspring. Fruit flies have been used in studies as diverse as alcoholism, learning and behavior, ecology and evolution, human disease and the development of new pharmaceuticals. It is fair to say that fruit flies have revolutionized biological science.

Fruit flies or *Drosophila melanogaster* live in almost all temperate regions of the world. The only aspects limiting the habitats is temperature and availability of water, as the scientific name actually means "lover of dew", implying that the species requires a moist environment. The common fruit fly is normally tan, and 3 mm in length and 2 mm in width. It has a rounded head with large, red compound eyes; three smaller simple eyes, and a short antenna. The female is slightly larger than the male, with black stripes on the dorsal surface of the abdomen used to determine the sex of an individual. Males have a greater amount of pigmentation concentrated at the posterior end of the abdomen. Reproduction is rapid, a single pair of flies can produce litters numbering in the hundreds within a few weeks, and the offspring become sexually mature with one week. As the name implies fruit flies live primarily on plant material, adults thrive on rotting plants and fruits, where eggs are laid.

HISTORY

The Drosophila research story begins in the early years of the 20th century. However, the first wave of fruit fly immigrants had arrived years before at ports in the Caribbean, carried across the Atlantic in slave ships from Africa and southern Europe. In the 1870s, in the immediate aftermath of the Civil War, the burgeoning trade in rum, sugar, bananas, and other fresh fruits delivered them north to Boston, New York, Philadelphia and other flourishing cities on the east coast. In the early 1900s, fruit flies were one among many animals used in the laboratory. Those early years were also witness to an explosion in experimental biology. In 1907, Thomas H. Morgan, a zoologist at Columbia University, began using the insect to substantiate the chromosomal theory of inheritance. (That chromosomes are the carriers of genetic information.) In early May 1910, while routinely examining his fly collection, Morgan spotted a fly with two white eyes instead of the usual red variety. The white eyed fly was a new mutant. In subsequent breeding the white eyed characteristic had disappeared, as Mendel would have predicted. Later generations, began to appear with, some with white eyes, in the ratio conforming to previously determined expectations. From this and subsequent work, Morgan conjured up a compelling amalgam of hereditary ideas. Life, for both Morgan and the fly would never be the same. His accomplishments led to his winning the Nobel Prize in 1933. In 1913, Alfred Sturtevant, a student of Morgan, created the first genetic maps using *D. melanogaster.*

As mentioned above, fruit flies have proven themselves as pioneers in genetic research. They also produce immense quantities of saliva, and the chromosomes inside the salivary glands are huge, a thousand times thicker than normal. Each chromosome is made up of many parallel strands of DNA that have failed to separate. Chemical staining of these super sized chromosomes reveals dark horizontal bands along their length, distinct landmarks corresponding to the positions of specific genes. These super sized chromosomes are similar to biological barcodes that gave biologists the first glimpse of genetic differences between individuals and populations.

Inebriated fruit flies have been used to help scientists find potential drug targets for alcoholism. This finding provides a crucial explanation of why some people seem to tolerate alcohol better than others. The discovery also sheds new light on many of the negative aspects of drinking, such as liver damage.

Fruit flies may seem to be unlikely heroes in the battle against drug abuse, but new research suggests that these insects could claim that role. Scientists have found that fruit flies can be used as a simpler and more convenient animal model for studying the effects of cocaine and drugs like amphetamine and methylphenidate on the brain.

Fortunately, the relationship between fruit fly and human genes are so close that sequences of newly discovered human genes, including those implicated in disease, can often be matched against their fly counterparts. This provides a lead toward the function of the human gene and could help in the development of new and effective drugs or other forms of treatment in the future. Such research has been responsible for a host of biological discoveries that began with the humble fruit fly more than a century ago. One wonders what the remainder of this century will unveil.

The Giraffe

—Magnificent but endangered!

According to a recent article in the Smithsonian magazine, giraffes are disappearing from Africa before scientists can even begin to understand them. Losing these magnificent animals to poachers would be a tragedy for a number of reasons in addition to the fact that people love watching them. The giraffe's anatomy is unique, unlike any other inhabitant on earth and the extremely long neck and knobby legs combine to elevate its stature as the tallest terrestrial animal.

Physiologists, anatomists and clinicians have also been intrigued by the challenges presented by the anatomy on three major systems: the cardiovascular which maintains blood pressure stability; the musculoskeletal, to support a vertically elongated body mass; and the nervous system to rapidly relay signally over long neural networks. It is difficult to know just what to make of giraffes, they shuffle like a camel (right legs forward, then left legs), but run like a rabbit (hind legs forward, then front legs). The giraffe bellows, hisses and moans and makes flute-like sounds in the wild, and in captivity hums in the dark. During courtship, males emit loud coughs. A giraffe naps with its head aloft, but sleeps like a swan with its head on its haunches.

The name "giraffe" has its earliest known origins in the Arabic word *zarafah,* translated as "fast-walker." Giraffes were once common throughout Africa, but their range has continuously diminished. Unfortunately, only about 300 West African giraffes are known to exist and other subspecies are also endangered.

Giraffes were new and exciting to people who had never seen them before, but not to those areas of Africa where they were killed for food even before guns were introduced to the continent. A dead giraffe offered a wealth of opportunity for fresh meat, the long bones contained much marrow and used for fertilizer, and the leg tendons could be used for sewing, or as guitar or bow strings. The hides were made into pots, buckets, drum covers, whips, sandals or shields. Long tails were especially valued for switches or ornaments, the hairs used as threads by Masai women to sew beads onto clothing. The tails are also used for marriage dowries.

Fully grown giraffes stand from 14 to 18 feet tall, with males taller than females. The average weight is about 2600 pounds for an adult male, and 1825 for a female. Their life span is unusually long compared to other ruminants, up to 25 years in the wild. (Ruminants have four chambered stomachs and regurgitate their food.) Giraffes have large bulging eyes which provide good all-around vision from the great height, and the senses of hearing and smell are acute. Giraffes can close their muscular nostrils to protect against sandstorms and ants. The lips, tongue (18 inches long) and inside of the mouth are covered in papillae to protect against thorns. Their thick saliva helps coat any thorns they may swallow.

As mentioned earlier, the most obvious giraffe feature is the extremely long neck, which varies from 6 to 8 feet. The reason is the result of a disproportionate lengthening of the cervical vertebrae, and not from the additional vertebrae. Birth is another oddity. A mother giraffe stands when it occurs and the baby precipitously drops five or more feet to the ground. Fortunately the new born is relaxed and unlikely to be injured.

The circulatory system of the giraffe has a number of variations due to its height. The turbocharged heart, which can weigh more than 25 pounds measures about 2 feet long and must generate approximately double the blood pressure required for a human to maintain blood flow to the brain. As such, the heart walls are about 3 inches thick. And giraffes have unusually high heart rates for their size, at about 150 beats per minute and the blood pressure is two and half times that of a human. Giraffes grow more hypertensive as they age. Fortunately the blood vessels in the lower legs are greatly thickened to withstand the increased pressure due to the weight of fluid pressing down on them. To solve this problem, the skin of the lower legs is stiff and tight, preventing too much blood from pouring into them. The skin thus serves the same task as a compression stocking and has been used in a number of scientific experiments, such as to develop suits for astronauts and fighter pilots.

Another strange characteristic is the giraffe's distinctive and strong aroma which can be detected over a considerable distance, estimated to be at least 250 meters or more than 800 feet. Two highly odoriferous compounds found in giraffe hair appear to be responsible. The scent may also serve a sexual function, as males smell stronger than females. Giraffes also have an unusual diet that includes toxic plants, shrubs and fruit and an appetite that includes eating up to 75 pounds daily.

Fortunately, there are institutions like the Giraffe Conservation Foundation in Namibia and others who endeavor to save these magnificent and majestic animals. Let's hope they are not too late.

Hand washing

—A New Year's Resolution worth keeping!

Many of us will consider resolutions for 2020, promises likely to be unattainable or short lived. Common ones relate to good health, including weight loss, a better diet, more frequent physical exams, or additional exercise. Perhaps the best one of all, and more apt to be followed is to consider washing your hands more often. Practicing good hand washing techniques is one of the easiest and most effective ways of preventing illnesses from spreading. This is particularly important in the workplace where large groups of people can catch the same infection. Below are some of the infections and diseases which can be spread by not thoroughly washing your hands. This information is available from the Centers for Disease Control and the World Health Organization.

Noroviruses

Norovirus is the most common cause of viral gastroenteritis in humans and it can affect people of all ages. It is transmitted when people don't wash their hands and worryingly, they can spread very quickly within large groups of people in close quarters. This is why when one person gets ill, entire households or offices often catch it too.

The best way to stop noroviruses from spreading or occurring in the first place is to wash your hands thoroughly after using the bathroom, before preparing food and to avoid touching your nose and mouth.

Airborne illnesses

Respiratory illnesses are usually spread via droplets which are breathed, sneezed or coughed into the air by someone who has the illness. While sneezing and coughing help to spread illnesses, poor hand washing techniques are a big culprit as well.

Common respiratory illnesses caused by poor hand hygiene include the common cold, influenza, chicken pox and meningitis.

Nosocomial infections

We often hear of infections being transmitted in hospitals and this is often the result of staff and patients not washing their hands. Naturally there's a huge amount of infections present in hospitals and if staff don't wash their hands between seeing patients or if people with an infection aren't practicing good hand hygiene, they can very easily pass their illness onto others.

Some of the most common nosocomial infections which can be spread by germs and bacteria on our hands include MRSA and E.coli.

Hepatitis A

Hepatitis A is a viral infection which can cause severe symptoms including problems with the liver, jaundice, abdominal pain, fever and fatigue. It's often spread via food which has been contaminated by people preparing it who haven't washed their hands after using the bathroom.

One of the biggest problems with not washing our hands after using the bathroom, according to the Mayo Clinic, is that throughout the day we touch many things. Other people then touch these things and when we then touch our nose or mouth, we pick up infections. As well as this, if you prepare foods with dirty hands, people can catch infections by eating what you have made.

When to wash your hands

As you touch people, surfaces and objects throughout the day, you accumulate germs on your hands. You can infect yourself with these germs by touching your eyes, nose or mouth, or spread them to others. Although it's impossible to keep your hands germ-free, washing your hands frequently can help limit the transfer of bacteria, viruses and other microbes.

Always wash your hands before:

- Preparing food or eating
- Treating wounds or caring for a sick person
- Inserting or removing contact lenses

Always wash your hands after:

- Preparing food
- Using the toilet, changing a diaper or cleaning up a child who has used the toilet
- Touching an animal, animal feed or animal waste
- Blowing your nose, coughing or sneezing
- Treating wounds or caring for a sick person
- Handling garbage
- Handling pet food or pet treats

Also, it goes without saying that you should wash your hands when they are visibly dirty.

How to wash your hands

It's generally best to wash your hands with soap and water. Over-the-counter antibacterial soaps are no more effective at killing germs than is regular soap.

Follow these steps:

- Wet your hands with clean, running water—either warm or cold.
- Apply soap and lather well.
- Rub your hands vigorously for at least 20 seconds. Remember to scrub all surfaces, including the backs of your hands, wrists, between your fingers and under your fingernails.
- Rinse well.

Dry your hands with a clean towel or air-dry them. Help children stay healthy by encouraging them to wash their hands frequently. Wash your hands with your child to show him or her how it's done. To prevent rushing, suggest washing hands for as long as it takes to sing the "Happy Birthday" song twice. A recent issue of the British Medical Journal recommends the popular nursery rhyme "Brother John" or "Frere Jacques." New words are added to teach the proper technique.

Final thoughts

Hand-washing offers great rewards in terms of preventing illness. Adopting this habit as a New Year's resolution can play a major role in protecting your health.

Happiness

—*Something we all desire.*

ONLY ONE OF the seven dwarfs is Happy, which humorously tells us that many people are not. Add that to the fact this country is not high on the list of nations where people profess to be happy. The U.S. currently ranks 18th worldwide in measures of happiness even though the Declaration of Independence affirms happiness as one of our most cherished goals. Where you live can affect your well being as does economic and social factors. It should not be a surprise therefore that Time magazine devoted an entire recent issue to various methods designed to achieve happiness—a subject that should be of interest to all of us. Factors that influence happiness include social support, years of healthy life expectancy, GDP per capita income, generosity, and freedom to make life decisions. In a recent editorial in the New York Times, the author provided another solution, he suggested that to be happy you should think like an old person. Older people report higher levels of contentment or well being than teen agers and young adults. It is a paradox of old age that instead of feeling worse about their age, they feel better. I would guess, however, that most everyone would prefer not to wait that long. There are other methods to consider and even an extremely complex mathematical formula for happiness.

Being mindful and meditating appear to be strategies designed to bring us joy, and while many people have a biased view of meditation, it is becoming a mainstream approach of reducing stress and improving mental well-being. Both are instrumental in making us happier. Mindfulness, living in the present moment, allows an individual to become more highly productive and possibly improve relationships with others—one of the predictors of success and happiness. Mindfulness can enhance pleasure, whether it is emotional or sensual (food or touch or sound). Being mindful allows us to savor the sensation or experience completely, it extends pleasurable experiences all the more and makes experiences more satisfying.

According to Psychology Today, religion can be a path to happiness. Religion is related to well being, and reasons include a sense of belonging, a sense of meaning in life, and a greater ability to exercise self control. One study found that people are more satisfied with their lives when they go to church, mosque or synagogue, because they build a social network within their congregation.

The Time magazine article noted that to experience more happiness each day you don't have to overhaul your life or spend a fortune. Instead buy yourself bliss by spending more time on social bonding rather than possessions. Second, is to bask whatever is great about yourself rather than try to fix what is not. Third, is to be generous with your time and money. Giving to charity brings more happiness than spending on yourself. Another factor is becoming more grateful for the blessings you have.

Gratitude allows people to report feeling content. One surprise is the mere act of smiling. It can cheer you up, although the reasons have yet to be discovered.

In a World Value Survey about general happiness that included more than 350,000 people in fifty countries, the following five factors in descending order have the strongest influence on our overall well being. They included family relationships and our close private life; our financial situation; our work, as something which gives our life purpose; community and friends as a source of trust and belonging; and our health, especially for those suffering from severe illnesses, and most extremely in cases of mental disorders. As indicated, family is most important in happiness. For example, going through a divorce has double the impact on one's level of happiness as losing 30 percent of one's income and married people have been shown to live longer and to enjoy healthier lives. The survey tells us that true companionship and a sense of belonging are two of the most positive things that people can experience in our lives. Despite being in 18th place, we are fortunate to live in the United States, as people who live in stable and peaceful countries are reported to be happier than those living in restricted societies. If you live in the United States, South Dakota and Vermont were tied for the nation's highest level of well being in 2017. West Virginia reported the lowest level of well being for the ninth year in a row.

Those of us with personal values are also shown to be happier, that includes have a positive philosophy of life. People who value their lives and appreciate what they have are known to be happier.

Harvey Washington Wiley and the Pure Food and Drug Act

THERE HAVE BEEN a number of celebrities working for the Food and Drug Administration, but I would venture that from the aspect of good health no one matches the accomplishments of Dr. Harvey Washington Wiley. His federal governmental and private services in the early 1900s were instrumental in changing the law to diminish false labeling for drugs and for keeping our food supply pure, safe and free from toxic additives and other contamination. Prior to his research and activism, anything manufactured by U. S. companies was suspect and many European countries threatened or refused to import food grown or packaged here. There were any number of dishonest food brokers—- we were becoming as infamous as those in England where pickles were made green by copper; rendered sharp with sulfurous acid; cream composed of rice powder or arrowroot in milk; jams mixed of sugar, starch and clay, and colored with preparations of copper or lead; and catsup formed of the dregs of distilled vinegar with a decoction of the outer green husk of walnuts, and seasoned with all spice. Fury over the danger posed by adulterated foods resulted in the passage in 1860 of Britain's Act for Preventing Adulteration in Food and Drink. Business interests managed to limit the fine for poisoning food to a mere £5, but at least it was a precedent. There was no such law in the United States. In addition to the similar problems here, honey was rampantly adulterated with corn syrup and milk was being preserved by the presence of formaldehyde. (Formaldehyde is usually used to stop dead bodies from rotting.) Coffee might be largely sawdust, or wheat, beans, beets, peas, and dandelion seeds scorched black and ground to resemble the genuine article. Flour routinely contained crushed stone or gypsum as a cheap extender. There were other incidents as well and many products were deliberately mislabeled. The United States was well behind the rest of Europe in terms of legislation. Many reporters, congressmen, and crusading private citizens clamored for a new law, no one more adamant and persistent than Harvey Wiley.

Wiley's early history and career were recently documented by Deborah Blum in her fascinating new book, *The Poison Squad*. He was the sixth of seven children, born on April 16, 1844, in a log cabin on a small farm in Kent, Indiana, about a hundred miles northeast of the farm where Abraham Lincoln had grown up a few decades earlier.

Harvey Wiley was preparing for college when the Civil War broke out, and his parents, despite their antislavery stance, were determined that he continue with it. He enrolled at nearby Hanover College in 1863 but a year later decided he could no longer sit out the war. After joining the 137th Indiana Infantry, he was deployed to Tennessee and Alabama, where he guarded Union-held railroad lines and spent his spare hours studying anatomy, reciting daily from a textbook to a fellow soldier. After discharge he accepted an offer to teach chemistry in the Indianapolis public high schools and there began to appreciate the insights offered by that branch of science or, as he came to see it, the "nobility and magnitude" of chemistry. Realizing he had a passion for the rapidly advancing field, he went back to school yet again, this time to study chemistry at Harvard University, which—as was typical at the time—awarded him a bachelor of science degree after only a few months of study. In 1874 he accepted a position at Indiana's newly opened Purdue University as its first (and only) chemistry professor.

Wiley developed a reputation as the state's go-to scientist for analyzing virtually anything—from water quality to rocks to soil samples—and especially foodstuffs. This was accelerated by a working sabbatical in 1878 in the newly united German Empire, considered the global leader in chemical research. He studied at one of the empire's pioneering food-quality laboratories and attended lectures by world-renowned scientist August Wilhelm von Hofmann, who had been the first director of the Royal College of Chemistry.

Unlike Great Britain, the U.S. did not have anything resembling a Food or Drug Administration. The closest department for analyzing food and drink resided at the Department of Agriculture, established by President Abraham Lincoln in 1862. The department had little to do with enforcement until the director named Wiley as Chief Chemist in 1883, recruiting him from Purdue University. He spent the next 25 years fighting for pure food, drink and properly labeled drugs. His work laid the foundation for the first federal law, entitled the Pure Food and Drug Law of 1906. Newspapers referred to the Act as "Dr. Wiley's Law." It was designed to prevent the manufacture, sale, or transportation of adulterated or misbranded or poisonous or deleterious foods, drugs, medicines and liquors, and for regulating their traffic. Unfortunately President Theodore Roosevelt did not give Dr. Wiley the credit he deserved and refused to acknowledge his heroic efforts.

HM and HeLa

*—Letters you may wish
to learn more about!*

PERHAPS THERE IS nothing more irritating than the expanded use of acronyms—particularly those associated with the internet that appear to grow at an annoying rate. (Acronyms are shortened forms of words or phrases to speed up communication.) I also struggle when acronyms are used once in the beginning of a book or article and my mind must recollect its meaning further on in the text. Unfortunately, anyone who is unaware about acronyms is out of touch with the times. One acronym I do find interesting is HeLa, it stands for Henrietta Lacks and is the code name given to the world's first immortal human cancer cells. They came from cervical cells removed from Henrietta Lacks just months before she died at age 31. These cells became instrumental in developing the most important cell lines in medical research. According to Rebecca Skloot, in her book *The Immortal Life of Henrietta Lacks*, that "if you could pile all HeLa cells ever grown onto a scale, they would weigh more than 50 million metric tons—an inconceivable number given that an individual cell weighs almost nothing." Another scientist calculated that if you could lay all HeLa cells ever grown end to end, they would wrap around the

Earth at least three times, spanning more than 350 million feet. (Skloot's book was on the best seller list for years.) Before Henrietta died the surgeon took samples of her tumor and put them in a petri dish. Scientists had been trying to keep human cells alive in culture for decades, but all of the cells eventually died. Henrietta's cells were different; they reproduced an entire generation every twenty-four hours, and they never stopped. Her cells were part of research into the genes that cause cancer and those that suppress it; they have helped develop drugs for treating herpes, leukemia, influenza, hemophilia, and Parkinson's disease; and been used to study lactose digestion, sexually transmitted diseases, appendicitis, human longevity, mosquito mating, and the negative cellular effects of working in sewers.

I am also intrigued by the persistent use of initials in medical reporting, the result of the need to maintain a patient's privacy. One set of interest is HM which stands for Henry Molaison. He is known by thousands of psychology students. HM had been knocked down by a bicycle accident at the age of seven, and began to have minor seizures at 9, and major seizures at 16. By the time he was 27, he became incapacitated by his seizures, despite high doses of anticonvulsant medication. In 1953, he underwent a bilateral medial temporal-lobe resection, an experimental brain operation intended to alleviate the severe epilepsy he had faced since childhood. The procedure used for Henry had been previously carried in psychotic patients, and the surgery was performed with the approval of the patient and his family. The operation did control his seizures, but with an unanticipated and devastating consequence consequence—an extreme amnesia that robbed Henry of the ability to form new memories and, in doing so, determined the course of the rest of his life.

For almost six decades, the scientists who studied Henry kept his name hidden away. When they wrote about him they were always careful not to reveal too much, for fear that outsiders might find him, and they were successful. There wasn't a single paper, out of the hundreds that chronicled in great detail the experiments performed on Henry during the fifty-five years between his operation and his death, that contained anything but the vaguest biographical information about Henry himself. Work with HM has established fundamental principles about how the memory functions are organized in the brain. As mentioned above, HM is likely the most studied individual in the history of neuroscience. Interest in his case can be attributed to a number of factors, including the unusual purity and severity of the memory impairment, its stability, and its well-described anatomical basis. Because HM was the first well studied patient with amnesia, he became the yardstick against which other patients with memory impairment would be compared. The study of HM established the key principles about how memory is organized that continue to guide the discipline.

There are two very interesting books that describe Henry's unique life. One written by Suzanne Corkin is entitled *Permanent Present Tense*, the other authored by Luke Dittrich is simply, *Patient H.M.*

Final thoughts

Progress in medicine is indebted to countless individuals, but much of the credit must be shared with both Henry Molaison and Henrietta Lacks. They achieved fame under extremely different circumstances. Henrietta was not aware of her contribution as she had longed passed away, whereas Henry had lost the ability to recall his.

HIBERNATION

—A bear's secret to good health.

As STATED IN the New York Times, scientists have been puzzled for decades over the evolutionary ability that allows bears and other hibernating animals to lie still through the winter, forgoing food and water, and yet emerge with their health intact come spring. Hibernation is one of the three major seasons in the life of a bear, it lasts from January until spring, then there is the active season, beginning in May, followed by a period of intense eating in late September. It is defined as a specialized seasonal reduction of metabolism concurrent with the environmental pressure of food inavailability and low environmental temperatures. The definition applies to both bears and certain other small mammals. Some bears exhibit dormancy for up to seven months. During hibernation the bear's metabolic and heart rate drop significantly. The resting heart rate drops from 40 beats per minute to as low as 8-10 beats per minute during hibernation. Bears do not defecate or urinate. The animal becomes resistant to insulin but does not suffer from fluctuations in its blood sugar levels. Even though the amount of nitrogen in the blood rises sharply, there is no damage to the kidneys or liver. Platelets in the bear's blood become less sticky, acting as a natural blood thinner, to counteract blood clots that could form during long periods of immobility. The bear's metabolism drops 25 percent of its normal rate and their kidneys actually stop functioning. If you or I would experience the same conditions under a similar time table we could end up with diabetes, obesity, bone loss, and atrophied muscles. If researchers could better understand the mechanism for hibernation, it may be possible to develop new drugs or medical treatments, an example would be helping astronauts survive long spaceflights or leading to drugs to treat diabetes or obesity. The physiology of hibernation might also be applicable to organ transplants. A waiting kidney or liver can be preserved in cold solutions for 24 hours, but can't be used after that. A heart or lung is only viable for four to six hours.

There has been a remarkable increase in hibernation research in the past few years, much directed to scientists studying obesity, which has become epidemic in the United States. Obesity in humans is associated with resistance to insulin, a hormone that regulates glucose in the blood, and Type 2 diabetes. According to one study, the bear's handling of insulin appears to vary with the seasons, with resistance increasing during hibernation and sensitivity increasing in the summer. Strangely, obese bears are healthier and more reproductively fit than expected, and have advantages counterintuitive to human biology. In one study, researchers took samples from the liver, fat and muscle of six captive grizzly bears at three times during the year. In the lab, they analyzed the DNA to understand the changes that occur in the cells over the course of a year. They found that the effect of hibernation on each tissue is different which means that hibernation is not just as simple as hibernating and not hibernating, as there are transitional events happening throughout the year. Fatty tissue changed the most, whereas muscle tissue barely changed. The muscle cells remained active through the hibernation period, which might help explain why those tissues do not atrophy (decrease in size).

The most recent more sophisticated study published in Communication Biology from Washington State University used RNA sequencing to reveal tissue and seasonal changes occurring in grizzly bears. Comparing hibernation to other seasons, bear adipose tissue has a greater number of expressed genes than liver or skeletal muscle. Hibernation is characterized by reduced expression of genes associated with insulin signaling, muscle protein degradation, and urea production. Across all tissues there is a subset of shared differentially expressed genes, some of which are uncharacterized, that together may reflect a common regulatory mechanism. The gene families could be useful for developing novel therapeutics to treat human and animal diseases.

Other animals hibernate, too, like mountain pygmy possums in Australia, ground squirrels in North America grasslands and various species of bats. Hibernating ground squirrels use melatonin, a potent antioxidant, to protect the cells when blood flow increases after months of inactivity. This information may be used to treat hemorrhagic shock to reduce damage to tissues when blood supply returns. A team of scientists has recently identified a potential drug to use in humans to treat brain damage caused by strokes using information derived from studying hibernating ground squirrels. Nearly 800,000 Americans experience a stroke every year.

FINAL THOUGHTS

According to Bill Bryson in his new book, *The Body*, bears , the most famous of wintry slumberers, don't actually hibernate. Real hibernation involves profound unconsciousness, and a dramatic fall in body temperature—to around 32 degrees. By this definition bears don't hibernate because their body temperature is near normal and they are easily aroused. Their winter sleeps are more accurately called a state of torpor.

Our microbiota

—An essential companion!

It is almost impossible to believe but there are approximately 100 trillion micro-organisms (most of them bacteria, but also viruses, fungi and protozoa) living in the human gastrointestinal tract.

This collection or community of organisms is called the microbiota. The collection of the genomes of the microbiota is defined as the microbiome and consists of over three million genes which produce thousands of metabolites. They replace many of the functions of each of us and influence our fitness, phenotype (observable characteristics or traits), and health. The microbiota is part of the human ecosystem.

According to one author, the armies of bacteria that sneak into our bodies the moment we are born are the "primal illegal immigrants." Most are industrious and friendly, minding their own business in tight-knit, long-lived communities, doing the grunt biochemical work we all rely on to stay alive. The ecosystem forms at birth, but the human-microbe alliance begins months before. Midway through pregnancy, a hormonal shift directs the cells lining the vagina to begin stockpiling sugary glycogen, the favorite food of sausage-shaped bacteria called lactobacilli. By fermenting the sugar into lactic acid, these bacteria lower the pH of the vagina to levels that discourage the growth of potentially dangerous invaders.

The infant mouth's first inoculation of bacteria includes a generous sampling of the lactobacilli present in the mother's birth canal. With the first gulp of breast milk, these lactobacilli are joined by millions of bifidobacteria, a related group of acid-producing microbes. The source of these bacteria are the mother's nipples, where the bacteria appear during the eighth month of pregnancy. Bifidobacteria secrete acids and antibiotic chemicals which repel potentially dangerous organisms including *Staphylococcus aureus*. Bifidobacteria and lactobacilli are soon joined with acid-tolerant *Streptococcus salivarius* that appear on a baby's tongue during the first day of life. Bifidobacteria are anaerobic (surviving without oxygen) rods which break down dietary carbohydrate and synthesize and excrete water-soluble vitamins. Their name is derived from the observation that they often exist in a Y-shaped form. These organisms predominate in the colons of breast-fed babies and account for up to 95% of all culturable bacteria and protect against infection. Strangely, they do not occur in such high numbers in adults. Several other streptococci along with one or more kinds of *Neisseria* bacteria settle in during the first week. The vast majority emanate from the mother's mouth, which is always within reach of a nursing baby's fingers.

As the baby begins nursing or drinking formula, the bacterial population inside the mouth increases. These bacteria consume enough oxygen, creating a zone where anaerobic bacteria can thrive. By the time the baby is 2 months old, a microscopic close-up of the gums will reveal clusters and chains of bacteria and fungi. Another wave of bacteria arrive when the first teeth appear. The first is *Streptococcus sanguis* followed by *Streptococcus mutans*. By middle childhood, the diversity inside the mouth surpasses a hundred species, and their total number is greater than 10 billion. Bacteria also settle in the nasal cavities, which are connected to the mouth via the upper respiratory tract. The bacteria eventually lodge in the intestinal tract. In the small intestine, incoming microbes engage the infant's dormant immune system.

When the child grows up to become an adult, his or her intestine is home to an almost inconceivable number of microorganisms. The size of the population—up to 100 trillion (a trillion seconds in time would be 32,000 years)—far exceeds all other microbial communities associated with body's surfaces and more than 10 times greater than the total number of our somatic and germ cells. (There is a significant variation in both the total number of bacteria and the composition of the bacterial flora in different body regions.) With aging, environmental factors related to diet, drugs, pesticides, continue to affect the composition of our microbiota. Since humans depend on their microbiota for various essential services, a person should really be considered a superorganism, consisting of his or her own cells and those of all the bacteria. A superorganism could also be described as colonies of individuals tightly knit by cooperation, complex communication and division of labor.

We are indeed fortunate, humans are not inherently endowed with a healthy immune or digestive system. Fortunately, our intestinal tract, which includes our inhabitants (microbiome), provides us with genetic and metabolic attributes we have not been required to evolve on our own, including the ability to harvest otherwise inaccessible nutrients and to modify host immune reactivity. There is still a lot to learn, however, scientists continue the effort to understand how microbes colonize the gut and to identify the genes that help microbes pass through the stomach's harsh environment and survive in the lower gastrointestinal tract.

Human Errors

IF YOU BELIEVE the theories in Charles Darwin's book, On the Origin of Species, natural selection allows nature to determine which species survive and which will perish. Natural selection implies that random variation, which is key to survival, allows survivors to pass along whatever advantages they may have to their offspring. *Homo sapiens*, the genus and species we humans belong to. This process now has been ongoing for more than 200,000 years which would appear to be enough time to develop our human bodies to function smoothly and effectively.

Every cell, every protein, and every letter in our DNA code has also been subjected to the harshness of natural selection over the fullness of evolutionary time. All of that time and all of that selection has resulted in a body form that should be fantastically robust, strong, resilient, clever, and mostly successful in the great rat race of life. But it is not perfect. For example, according to Howard Lents' new book, *Human Errors*, "We have retinas that face backward, the stump of a tail, and way too many bones in our wrists. We must find vitamins and nutrients in our diets that other animals simply make for themselves. We are poorly equipped to survive in the climates in which we now live. We have nerves that take bizarre paths, muscles that attach to nothing, and lymph nodes that do more harm than good. Our genomes are filled with genes that don't work, chromosomes that break, and viral carcasses from past infections. We have brains that play tricks on us, cognitive biases and prejudices, and a tendency to kill one another in large numbers."

Our flaws illuminate not only our evolutionary past but also our present and future. One of the major problems is our upright posture. It was adapted from a body plan that had mammals walking on all fours. This tinkering aided our early ancestors as standing on our own two feet promoted tool use and enhanced intelligence. On the other hand our backbone has since adapted somewhat to the awkward change, the lower vertebrae have grown bigger to cope with the increased vertical pressure, and our spine is curved a bit to keep us from toppling over. These fixes have resulted into a number of orthopedic problems, particularly as we age. Bones lose minerals as we continue to grow older, the demineralization makes bones susceptible to fracture and osteoporosis. Spinal disks can slip, rupture or bulge. Muscles continue to lose mass and tone and weakness contributes to lower back pain. Veins in the legs are more prone to varicosity, they become enlarged and twisted. Valves that should snap shut between heartbeats malfunction, causing blood to pool. Severe varicose veins can lead to swelling, pain and on rare occasions, life threatening blood clots. Knee joints in particular, are somewhat fragile and complicated, cartilage is worn away with repetitive use, causing bones to grind against each other. The resulting pains may be exacerbated by osteoarthritis. Another orthopedic problem stems from humans having a relatively short rib cage, which does not fully protect our internal organs.

According to an early article in Scientific American, there are number of anatomical fixes needed to improve our overall well being. Examples include having a forward-tilting upper torso to relieve pressure on vertebrae. A curved neck with enlarged vertebrae would counter balance our tilted torso. Leg veins should have more check valves to combat varicose veins. Having knees that bend backward would make bones less likely to grind and deteriorate. Thicker vertebral disks would resist destructive pressures and thicker bones would protect against breakage during falls.

Various parts of the face, head and neck are other considerations with aging or before. For the latter, we are unfortunate to have a common upper passageway for food and air. When food travels toward the esophagus, a flap like tab of cartilage (the epiglottis) closes off the trachea, or windpipe. A progressive loss of muscle tone with age decreases the tightness of the seal, raising the risk of inhaling food or drink. There is a weak link between the retina and the back of the eye. This frail connection exists in part because the optic nerve, which carries visual signals from the retina to the brain, connects to the retina only from the inside of the eye, not from the back. Fixes could include attaching the optic nerve to the back of retina, and raising the trachea would help food and drink bypass the windpipe more effectively. Unfortunately, it is unlikely that evolution will fix our inherent flaws and future generations must continue to live with them.

Poliomyelitis (Polio)

—Still hanging around.

When I was young, many years ago, my parents did not allow my brother or me to go swimming at our local pool because it was thought to be the source of infantile paralysis. (Infantile paralysis is an old synonym for poliomyelitis, an acute and devastating viral disease.) My parents were not alone, it was common knowledge that children could be exposed to the virus there. Later a study conducted in 1946 showed that chlorine was actually one of the few known chemicals that could inactivate the virus.

Poliomyelitis had been frightening the country beginning in the early 1900s and continued until polio vaccine was developed. (It is derived from *polios;* meaning gray, and myelitis, indicating inflammation of the spinal cord.) The first major documented polio outbreak in the United States occurred in Rutland County, Vermont in 1894. Eighteen deaths and 132 cases of permanent paralysis were reported. Charles Caverly, MD, noted the appearance of acute nervous system disease and was one of the first physicians to recognize that polio could occur with or without paralysis. The contagious nature of the disease was established in 1905. Beginning from about 1916, a polio epidemic appeared at least one part of the country, with the most serious cases reported in the 1940s and 50s. In the epidemic of 1949, 2,720 deaths from the disease occurred in the United States and 42,173 cases were reported. Polio is a debilitating virus (poliomyelitis) that has struck fear in the hearts of parents for centuries. It causes severe muscle weakness and paralysis, often affecting the spinal cord, difficulty breathing and sometimes death.

The virus is typically contracted by ingestion, usually when the individual's hands have been exposed to the feces of someone who is infected and subsequent swallowing of the virus. Polio is very contagious and spreads through person to person contact through droplets from a sneeze or cough. An infected person may spread the virus to others immediately before and about 1 to 2 weeks after symptoms occur The virus can live in an infected person's feces for many weeks and can contaminate food and water in unsanitary conditions. It thrives in areas where living conditions are unclean and hygiene is poor. Man is the only source for poliovirus. The virus enters the mouth and multiplies in lymphoid tissues in the pharynx and intestines.

What appears to be the polio virus can be traced back to 1580 B.C. Egyptian carvings and paintings show people with withered limbs walking with canes. These people appear to be healthy other than their affected limbs. They also show young children walking with the aid of canes. While the virus does affect adults, it is more common in young children and infants. When it infects an older individual such as an older child, teenager or adult, the chances of paralysis and severity of the effects of the virus are much more common. There are three types of polio viruses, types 1, 2 and 3. Type 1 is the most virulent and common.

The world is indeed fortunate following the introduction of a vaccine designed to prevent polio. Jonas Salk became a national hero when he allayed the fear of the dreaded disease following approval of his injectable vaccine in 1955. Although it was the first, it was not be the last; Albert Sabin introduced an oral vaccine in the 1960s that replaced Salk's. Both Salk and Sabin vaccines are trivalent, that is active against all three virus types.

In 1988, the World Health Organization (WHO) set its sights on eradicating polio, this came after success against smallpox. Since then, the incidence of polio has decreased from an estimated 350,000 annual cases to 1315 in 2007. Indigenous type -1 and type-3 have been eliminated worldwide from all but four countries (Afghanistan, India, Nigeria and Pakistan). It was thought that type-2 had been eradicated, but just last year type -2 vaccine derived virus is still circulating and last year caused polio in Syria and the Democratic Republic of Congo. This year cases emerged in Nigeria, Niger, and Somalia. Health officials worry that outbreaks in Somalia, in particular, may spread to neighboring countries. This is a setback for Africa, the last person on the continent paralyzed by the wild polio virus was a Nigerian child who contracted the disease in 2016. In 2017, cases caused by vaccine-derived viruses overtook, for the first time, those caused by the wild version.

They are still rare, but attracting more notice than those caused by the wild virus.

Fortunately, thanks to the vaccines, the United States has been polio-free since 1979, but it is still a threat in other countries and requires all of us to be vaccinated on schedule. People most at risk are those who never have been vaccinated, those who never receive all the recommended vaccine doses, and those traveling to areas that could put them at risk of acquiring polio. Unfortunately there is a growing number of countries with cases of the disease caused by the vaccine.

Human challenges

—How long has this been going on?

In the Netherlands, according to a recent report, four healthy volunteers took part in a controversial experiment in which *Schistosoma mansoni* parasites were permitted to burrow into a number of volunteer's arms. The parasite is a tiny waterborne worm species that causes schistosomiasis, a disease that sickens millions of people in Africa, the Middle East, and Latin America and kills thousands each year. The experiment is designed to use these patients to develop a vaccine to prevent the disease. Such research is not uncommon, other studies have been done with malaria, cholera, influenza, shigella, dengue, norovirus, tuberculosis, rhinovirus, typhoid and giardia. Human challenge studies, which only involve a few dozen volunteers, speed the process of deciding whether to pursue a promising treatment, which saves time and money. Moreover, tests that intentionally infect people can quickly and efficiently flag potential side effects. Human challenges date back to the 18th century and the first vaccine, when English physician Edward Jenner attempted to persuade the world that infecting a person with harmless cowpox could prevent the more dreaded disease, smallpox.

Self experimentation with drugs began as early as the 19th century, as reported by Friedrich Wilhelm Serturner, a German scientist. In 1803, he had numerous brushes with death as he experimented on himself to produce a drug that continues even today to be the ultimate painkiller—morphine. (Morphine is derived from opium.) His research was in fact the turning point in the history of drug studies. Serturner observed that some samples of opium could rapidly and completely dull pain and he theorized that opium contained a specific active ingredient responsible for this effect. In his quest to find that component he was the first to apply basic chemical analysis. He discovered the specific narcotic substance contained in opium, and named it "morphine" after Morpheus, the god of sleep. Once he isolated morphine, Serturner began testing it with cautious experiments on animals. Then he began to test it on himself and three young friends. They began ingesting a small dose, then continued to do so until each experienced a sharp stomach pain and a feeling that they were about to faint. Concerned with these symptoms, Serturner recognized that the drug was a poison and subsequently that they had swallowed about ten times the dose now recommended.

One other incident is worth mentioning. Max von Pettenkofer who lived in Munich, Germany in the late nineteenth century was convinced that it took more than cholera bacteria to cause the disease. To prove it, he swallowed one cubic centimeter of boullion laced with the organism derived from a patient who had died from cholera. The next day he began to experience abdominal colic with extensive gas pains and diarrhea which lasted almost a week. Fortunately he did not die, likely due to the fact that he had contracted an extremely mild case of the disease.

Intentionally infecting a human being with a deadly disease would not pass ethical standards in the civilized world today, but as recently as the early 20th century, Australian psychiatrist Julius Wagner-Jauregg won the 1927 Nobel Prize in Medicine, for injecting blood from people with malaria into patients with syphilis, in a effort to cure them from insanity and paralysis. The treatment was quite successful, and many thousands of victims were spared almost certain lingering death from the disease. However, with the advent of penicillin in the mid 1940s, it was obvious that malaria treatment would be slowly but inevitably replaced.

Many doctors like the aforementioned Pettenkofer, have challenged themselves with pathogens to prove the worth of their own experimental medicines or theories. Another example is Dr. Barry J. Marshall, who, along with Dr. J. Robin Warren, identified a bacterium now known as *Helicobacter pylori* in patients with inflamed stomachs and ulcers. As part of his research, Marshall swallowed a tube which was used for tests to document that he had neither gastritis nor an ulcer and was not silently harboring *H. pylori*. Then Marshall swallowed the bacteria, which led to gastritis. Although he did not continue the self-experiment long enough to produce an ulcer, his experiment provided strong indirect evidence that many ulcers result from *H. pylori* infection and are not caused by stress, as was previously thought. Marshall's role was one part of continuous research that led to antibiotic therapies that can cure gastritis and ulcers as well as the strong suspicion that chronic *H. pylori* can cause stomach cancer. (Anyone interested in the history of self experimentation should read Who Goes First? an excellent book published by the University of California Press.)

Today, of course, federal laws and regulations mandate human experimentation in large numbers of patients for medical research. Before a drug can be marketed in the United States it must pass through three stages for approval. The first involves tests of toxicity and safety on just a few human volunteers. Then the stages are enlarged in scope and participation.

Isaac Newton

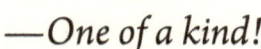

—*One of a kind!*

IN THE SPRING and summer of 1665 England suffered from an outbreak of Bubonic Plague, it spread from parish to parish until thousands died. Isaac Newton was attending University of Cambridge's Trinity College during that time and when the school closed he went home to continue his studies and research. Freed from the control of any professors, he was able to explore his ideas with even more curiosity, and take more intellectual risks than he had been able to at school. He read voraciously, and developed new theories about math and science, perfected telescope optics, discovered new information about the solar system, worked on developing the mathematics of calculus, and described the laws of light, motion and gravity. With this in mind, one would wonder what would happen if we closed our universities for one year told the students to study on their own, and then report each of their new ideas or discoveries. Odds are, the collective or individual results would not approach any of Newton's seminal contributions. In fairness, however, Newton was a genius who revolutionized science. He has been called the greatest mind of all time—his discoveries helped shape the world.

James Gleick, in his book about Newton, wrote the following: "He was born into a world of darkness, obscurity, and magic; led a strangely pure and obsessive life, lacking parents, lovers, and friends; quarreled bitterly with great men who crossed his path; veered at least once to the brink of madness; cloaked his work in secrecy; and yet discovered more of the essential core of human knowledge than anyone before or after. He was chief architect of the modern world."

Newton laid the foundation for the world we live in. Theoretical and applied science owes its thanks to Sir Isaac Newton for carving the paths so arrogantly strode. Because of his scientific breakthroughs, other esteemed scientists, such as Nikola Tesla and Thomas Edison, were able to complete the work that lay before them. Without Newton, none of that would have happened. Isaac Newton was a wonder of his day and age. According to Gleick and other biographers, "Newton stood aloof from other philosophers even after becoming a national icon—Sir Isaac, Master of the Mint, President of the Royal Society, his likeness engraved on medals, his discoveries exalted in verse."

Isaac Newton was born on January 4, 1643, in Woolsthorpe, Lincolnshire, England. The son of a farmer who died three months before he was born, Newton spent most of his early years with his maternal grandmother after his mother remarried. Fortunately his education was interrupted by a failed attempt to turn him into a farmer.

Although Isaac's early educational career got off to a slow

start when he was attending grammar school, his apprenticeship to an apothecary (a pharmacist) did spark an interest in science. Ironically, in the 1600s, apothecaries were doing much more scientific research (as we would understand the idea of "scientific research" today) than schools or universities were doing. While schools took almost all their information from "the classical authorities" like the works of Euclid, apothecaries actually had a reason to care about finding out new information – so that they could save more lives with their medicine.

At age 19, following grammar school, Newton enrolled in the University of Cambridge's Trinity College in 1661. He returned to Cambridge in 1667 and was elected a minor fellow. One year later he constructed the first reflecting telescope, and the following year received his Master of Arts degree and took over as Cambridge's prestigious Lucasian Professor of Mathematics. Asked to give a demonstration of his telescope to the Royal Society of London in 1671, he was elected to the society the following year. Years of research culminated with the publication of "Principia" in 1687, a landmark work that established the universal laws of motion and gravity. The book earned universal acclaim as one of the most important works of modern science. His second book, "Opticks," detailed his experiments to determine the properties of light. Through his experiments with refraction, Newton was the first to determine that white light was a composite of all the colors in the spectrum, and asserted that light was composed of particles rather than waves. Newton was an ardent student of history and religious doctrines, his writings on those subjects were compiled into multiple books published posthumously.

Newton is remembered as a transformative scholar, inventor and writer. His precise methodology gave birth to what is known as the scientific method. Although his theories of space time and gravity eventually were superseded by those of Albert Einstein, his work remains the bedrock on which modern physics was built.

Itching

—A diabolical, peculiar and often unsolvable problem!

There was a report of a patient with a severe itching problem on the scalp that resulted in prolonged and continuous scratching all the way through the skull and into her brain. The injury left the patient partially paralyzed. While this case is extremely rare, many of us have had been plagued with some condition where itching occurs. In fact, relentless itching has been described as among the most distressing of all physical sensations. While itching is technically merely a symptom, it affects a patient's quality of life. Moreover, mental processes are involved that help explain why it feels good to scratch. Within the past decade, there has been a flurry of research into what causes itching and how scratching helps to stop it. Along with brain imaging, studies have begun to look at gene activity and to map the signals that flow between cells in the skin and the immune system. It has been discovered that itching and scratching engage a number of brain areas. There is still much to learn, however.

Acute itching, such as from an insect bite, is often temporary and annoying and can be treated just by scratching. Chronic pruritus, on the other hand, is defined as an itch persisting for more than six weeks. It may involve the entire skin or only particular areas, such as the scalp, upper back, arms, or groin. Itching is an extraordinarily common symptom. All kinds of dermatological conditions can cause it: allergic reactions, bacterial or fungal infections, skin cancer, psoriasis, dandruff, scabies, lice, poison ivy, sun damage or just dry skin. Plenty of non skin conditions can also cause itching including hyperthyroidism, iron deficiency, liver disease, cancers like Hodgkin's lymphoma and psychogenic disorders. The number one cause of chronic itch is eczema. A recent survey shows that about 1.8 million Americans report suffering from the chronic form of eczema, about 6 percent of the population in the United States. In the Western hemisphere, around 20 percent of children have chronic eczema. Recent research indicates that the underlying cause for the inflammation is a toxin-producing bacteria found on the surface of the skin. Ninety percent of patients with eczema have exceedingly high numbers of bacteria on their inflamed skin.

Itching worsens at night in many systemic and skin diseases. Up to 65 percent of patients with inflammatory skin conditions including psoriasis, atopic dermatitis and other unknown causes have reported increased itching at night. Changes in skin physiology, such as temperature and barrier function may also play a role. Itching has been reported to be aggravated by warm temperatures and it has been suggested that heat can increase itch sensation by its effect on nerve endings. Although there are several types of temporary remedies available—both over-the-counter and prescription—-that may relieve non specific pruritus, the lack of treatments for night time itching is both astonishing and alarming.

Scratching is a natural response to an itch, and by definition, inseparable from it. The itch-scratch cycle is a complex phenomenon that involves sensory, motor and emotional components. It is well known that the urge to scratch can be remarkably intense, since the reward provided by scratching brings not only itch relief but associated feelings of pleasure as well. Itching and scratching engage brain areas involved not only in sensation, but also in mental processes that help explain why we scratch—- including motivation and reward, pleasure, craving and even addiction. What an itch turns on, a scratch turns off. Scratching may be the only way to relieve atopic eczema.

In chronic eczema, the more the patient scratches the more damage occurs to upper nerve layers of the skin barrier. These nerve fibers become overactivated so that scratching actually intensifies the perception of itch, causing greater itching. Scratching is also wide spread in the animal kingdom, although no one knows for sure why animals claw, bite or peck themselves.

Treatment of chronic pruritus should be directed to the underlying cause when possible. In the initial treatment of symptoms, the use of mild cleansers, topical anesthetics, and coolants may be helpful. Topical steroids, however, are the primary treatment because they help to reduce inflammation. It is also important to apply moisturizers to protect the upper layer of the skin. The reason is that water loss from the skin increases at night. Antihistamines that cause drowsiness may also be used, primarily to help a patient sleep. For some unknown reason, babies born in a home with a dog during pregnancy receive protection from allergic eczema that lasts until approximately age 10.

Currently there is no drug specifically designed to treat chronic itching, regardless of disease origin, approved by the Food and Drug Administration. One investigational one, however, appears promising but approval may be years away. For many patients, new treatments cannot come soon enough.

Knots

—A tied and true hobby!

*"Right over left, then left over right,
Makes the knot both tidy and tight."*
Reef knot mnemonic

I AM OFTEN on the alert for something that can occupy my mind and time that may be useful in the future. While browsing through the book store I came upon *The Handy Book of Knots* written by Randy Penn. It offered ten top reasons for learning to tie knots:

1. Knotting will help you in endeavors you already enjoy.
2. It will help you get involved in new activities.
3. Knotting is a useful skill in many crafts.
4. Tying cordage properly will make you look competent.
5. You will be able to manage ropes of different sizes and materials.
6. You will learn how to tie safer knots that won't untie under duress.

7. You'll react better if you need to tie knots in an emergency.
8. You'll find more uses for rope, like securing cargo to your bicycle or motorcycle.
9. You'll be able to rely on cordage to make a variety of repairs.
10. Most importantly, you'll learn how to do the job right.

From this description, it would appear that knot tying should be a required course in schools and universities. The more I read the more I became interested in the history, names, numbers, functions, and diversity of knots. There is even research carried out on how much force is required to tighten knots. This distinction helped sailors and others to choose certain knots over others to secure vessels.

History

Not much is known about knots and material used to tie them before recorded history, but both plant and animal materials were available for use in cordage. (Cordage is a general term referring to ropes and twine.) Numerous plants are made of strong fibers that provide structural strength as do a variety of animals. Animal hides could be cut into thin strips for tying materials. Early humans must have been inspired to tie their first knots by what they saw around them. Spider webs, bird nests and even the complex nature of many plants may have provided suggestions.

BACKGROUND

For anyone wishing to learn more about knots, there is a treasure house of information in *The Ashley Book of Knots*, first published in 1944 and still in print. It contains nearly 7000 illustrations of over 3000 knots. The book for which Ashley dedicated over a decade of research, writing and illustration, remains the essential guide to knot tying, and a historical archive of their uses. Ashley's goal was to document and reproduce every known knot, cataloguing them by their uses and designs, and offer instructions on how to recreate their specific ties.

In July 2017, the New Bedford Massachusetts Whaling Museum opened a special exhibit on Ashley's work entitled Thou Shall Knot. The museum composed this description: knots "are integral to the ships we sail, the clothes we wear, the hair we braid, the memories we keep, our colloquial expressions, the games we play, the shoes we tie, the presents we give, the fish we catch, and the social contacts that bind us."

NOMENCLATURE

The names given to knots provide clues about what they meant to our ancestors. One of the first things to notice is that some refer to professions. From Archer and Bell Ringer to Weaver, knots continue to be called by their namesakes—they played a key role in these trades. Some knots have multiple names, and one name can refer to many different knots. When a knot has many different names, it is an indication that, for whatever reason, the knot was significant enough to warrant such attention.

DIVERSITY

As noted by the description supplied above by the Whaling Museum, knots have a number of practical uses. There are so-called stopper knots or knots that are tied at the end of cord. Stopper knots have many uses, the knots allow the user to suspend an object or keep a line from running all the way out. Tying bends are used to join the ends of two ropes, to provide more length or make a needed connection. Loops can be used as a bowline or noose. Hitches can be used to secure a rope to rings, rails, posts, hooks, or other ropes. Lashing knots can secure a box or sleeping bag, or grouping a stack of items. There are also special knots for decoration purposes. There may be as many as 3800 different types.

MECHANICS

Researchers have analyzed the mechanical forces underpinning simple knots and carried out experiments to test how much force is required to tighten them. They discovered that the topology of a knot determines it mechanical forces. (Topology is the mathematical study of the properties preserved through deformations, twisting and stretchings of objects.) For the first time, scientists can predict the force needed to close knots. This may help understand something as simple as to how headphones become tangled, how to better tie shoe laces and how the configuration of knots can help in surgical procedures.

LEECHES

—Still used in medicine

WHEN I WAS much younger it was not uncommon to purchase leeches at the local drug store and use them to treat a black eye. Today modern medicine in the United States would generally find such use archaic and abhorrent. Not so in Russia, leeches are still widely prescribed there, about 10 million of them every year, in many cases as a low cost substitute for blood thinners like warfarin. In that country, a medicinal leech costs less than a dollar, and a typical treatment requires three to seven of these ravenous blood suckers. According to the New York Times, leech treatments take about 30 to 40 minutes, though the resulting wounds ooze blood for an additional six hours or so until the natural anticoagulant in leech venom wears off. In Russian medicine, the focus is on the leeches' venom—it is prescribed as a preventive treatment for stroke and heart disease at a fraction of the cost for drugs used as anticoagulants. (In the last few years the Food and Drug Administration (FDA) has approved three new oral blood thinner drugs, Pradaxa, Xarleto, and Eliquis. All three are primarily used to reduce the risk of atrial fibrillation.) Leeches are also used in Russia to treat glaucoma, prostatitis, high blood pressure and many more ailments. Patients are encouraged to use leeches in conjunction with standard drug treatments. Medical leeches sell for 90 cents each in Russia compared to 15 dollars or more here. Fortunately leeches can be used multiple times.

In developed countries leeches are creeping back into Western medicine, and in the U.S., 6000 leeches are used to remove excess blood from severed body parts that have been reattached.

Leeches actually have FDA approval and are classified as medical devices. However it should be noted that the medicinal leech has been used for eons. Their use was first recorded in 200 B.C., and according to historians, such therapy began even sooner in India. In the Napoleonic era, the Empress Josephine was treated with leeches after a fall. Napoleon's military surgeon advocated using thirty to fifty leeches at a time, enough to draw up to one and a half quarts of blood. (Leeches can consume over five times their body weight.) During that time it was estimated than more than 40 million leeches were used annually in France. In the late 1800s, leeching became so popular that the common preferred variety became an endangered species. The great advantage was that leeches could extract blood with very little pain and could be used on almost any part of the body. By the early part of the twentieth century, however, the use of leeches in medical treatment went into steep decline.

Leeches are closely related to earthworms and lugworms; they thrive in mountain lakes, desert oases and even polar oceans. There are 650 known species and unlike other worms leeches have a sucker at each end, one for feeding and one for hanging on. The medicinal leech (*Hirudo medicinalis*) has three jaws with 100 teeth in each. The specific value of leech therapy is related to the properties of its bite. The slimy creatures manufacture a wide portfolio of substances that help keep the blood flowing once they have attached themselves to a host. They don't just latch on, they pump out anticoagulants that prevent the wound they create from clotting too quickly. Leeches secrete hirudin, a direct thrombin inhibitor. Other substances produced by leeches include a vasodilator, hyaluronidase, and an anesthetic. Hyaluronidase is an enzyme which increases the permeability of leech saliva through human tissue and it exhibits antibiotic properties as well. The leech is also unique in that its digestive tract contains a single bacteria species, *Aeromonas veronii*. This organism may inhibit the proliferation of other bacteria.

In the late twentieth century there was a renewed interest in leeches by both the lay and scientific communities. This stemmed from clinical results in 1960 by two Yugoslavian surgeons. They described their use of leeches to relieve venous congestion caused by skin implants. The work was published in the British Journal of Plastic Surgery. Since then leeches have been used to repair grafted skin flaps and in the removal of congested venous blood in digital, scalp and external ear re-implantations. Leeches have also been studied where other treatments have failed to provide relief for the pain and inflammation of osteoarthritis. Other uses include breast reconstruction after a mastectomy, black eyes, and for reduction of postoperative swelling. Who knows—leeches may soon available from your local pharmacy.

Leprosy

—Not yet a thing of the past.

LEPROSY MAY BE the oldest recorded disease, it is mentioned 68 times in the Bible, (55 times in the Old Testament and 13 in the New), where most likely it meant a variety of infectious skin diseases. While the precise meaning of leprosy in both the Old and New Testament may be in dispute, it no doubt included the actual illness, now called Hansen's disease. The fact that leprosy has been with us so long is indicative of how persistent the causative agent is. And while leprosy is currently classified as rare, it still strikes some 200,000 people a year, most of them in Brazil, India, Mexico, the Pacific Islands and other Third World nations. Approximately 200 cases, however, occur annually in the United States almost all of which involve immigrants from developing countries who has settled in California, Hawaii and Texas. Perhaps, the presentation of the disease is what fascinates and frightens us most—- an infected patient may have grotesque skin lesions, and disfigurement, including clawed hands and feet , and a face that resembles a gargoyle. These bizarre symptoms led people to believe that leprosy was a curse, or a punishment from God. Lepers were considered dangerous and stigmatized: they had to wear special torn clothing, cover the lower part of their faces, cry out "Unclean! Unclean!, and ring bells to warn others of their approach. They also had to live in isolation outside the town walls. In modern times, treatment is confined to separate hospitals or live-in colonies called leprosariums. Despite its notoriety and antiquity, leprosy continues to surprise and confound researchers, and is still in the news.

HISTORY

Leprosy has terrified humanity from the beginning of recorded time—it was reported as early as 600 BC in Indian, China, and Egypt, and genomic studies of certain strains of the disease suggest that leprosy's parasitism in humans dates back more than 100,000 years. It is most likely that the leprosy bacilli started parasitic evolution in early hominids millions of years ago. The disease seems to have originated in Eastern Africa or the Near East and spread with successive human migrations. Europeans or North Africans introduced leprosy into West Africa and the Americas within the past 500 years.

ETIOLOGY AND CLINICAL TYPES

Leprosy is a chronic infection of the skin and nerves caused by *Mycobacterium leprae* and the newly discovered *M. lepromatosis.*). (Lepra is the Greek word for scaly.) Armauer Hansen, a Norwegian scientist, identified the agent responsible for the disease in 1873 but had no success in transmitting it to himself or to others.*

Mycobacterium leprae, a close relative of the tubercle

bacillus, is an obligate (incapable of adapting to different conditions), intracellular pathogen with a unique affinity for peripheral nerves, skin, and mucous membranes. Leprosy bacilli prefer the cooler temperatures that prevail close to the surface and explains why they seek refuge in testicles, earlobes, corneas, and nasal passages. In addition to the effects on the cornea, the organism also appears to numb the blink reflex. The blink reflex opens and closes the eyelids every 20 seconds or so to assure the eye stays lubricated. Without that lubrication, the surface of the cornea dries out and become susceptible to damage and ulceration.

The means of transmission is uncertain, but, like tuberculosis, the infection is thought to be spread by the respiratory route by sneezing or coughing because lepromatous patients harbor bacilli in their nasal passages. The bacteria accumulate principally in the extremities of the body where they reside with macrophages and infect the Schwann cells of the peripheral nervous system

Leprosy may occur at any age, although the peak age of onset is in the first, second and third decades. The most vulnerable persons are children who have prolonged contact with infected patients. There are three clinical types: tuberculoid, lepromatous and borderline. Patients with tuberculoid have one or a few hypopigmented, hypoesthetic macules with well defined borders. Peripheral nerves may be damaged and enlarged and frequently contiguous to skin lesions. Patients thus lose the ability to experience peripheral pain. Tuberculoid patients suffer from patches of dead skin, loss of sensation, nerve damage, but no extensive disfigurement. Many of the symptoms result from the body's furious autoimmune response to the foreign bacilli. Lepromatous leprosy is much more serious, patients develop symmetric skin nodules loaded with *M. leprae.*

CURRENT TREATMENT

Leprosy can be classified on the basis of clinical manifestations and skin smear results. The drugs generally used and recommended by the World Health Organization are a combination of rifampicin, clofazimine and dapsone for MB leprosy patients and rifampicin and dapsone for PB leprosy patients. Among these, rifampicin is the most important antileprosy drug and therefore is included in the treatment of both types of leprosy. Treatment with only one antileprosy drug will always result in drug resistance. Treatment with dapsone or any other antileprosy drug used as monotherapy should be considered as unethical practice. Multidrug therapy aims to effectively eliminate *M. leprae* in the shortest possible time to prevent resistance from occurring. Currently the duration of therapy is 12 months.

According to a recent letter from a leprosy researcher, the medical world still does not know exactly how leprosy is transmitted. Despite its longevity, leprosy continues to confound and surprise all who study its effects. We do know that the leprosy bacillus is remarkably poor at migrating between human hosts and that the organism fortunately dies quickly outside the body. Much, however, remains to be learned about a disease that is not yet a thing of the past.

LIGHTNING BUGS

―∞―

—Vanishing memories of youth!

ONE OF THE joys of growing up was chasing lightning bugs in the early evening during the summer and collecting them in glass milk bottles. This, of course, was many years ago, before cell phones and other digital technology, and children actually played outside of the house. Unfortunately there have been recent reports that one of the fond memories of my youth may be lost. Lightning bugs, more commonly called fireflies, are facing increasing threats from habitat loss, pesticides and pollution. Their demise would be a tragedy. There are three reasons likely to blame. The first may be loss of habitat. Mangroves, forests, wetlands, rice paddies and marshes are vanishing together with fireflies. All are important because they provide special conditions to complete the fireflies life cycle. Second is light pollution from cities, billboards, street lights and houses. Bright lights interfere with and outshine the mating signals fireflies use to attract females. Bright light also disrupts the feeding patterns of the females that glow to attract and eat males. The third is the persistent use of insecticides and pesticides. Additional threats include global warming and water pollution.

Fireflies have been eloquently described by Sara Lewis, in her book, *Silent Sparks,* as follows "Their resplendent displays change ordinary landscapes into places ethereal and otherworldly. Fireflies can transform a mountainside into a living cascade of light, a suburban lawn into a shimmering portal to another universe, a serene mangrove- lined river into a hypnotically pulsating disco."

Since their evolutionary origin some 297 million years ago, beetles have been highly successful; they represent 38% of known insect species. Fireflies rank among the most charismatic beetles, with distinctive bioluminescent courtship displays that make them a potential flagship group for insect conservation. With more than 2000 species worldwide, firefly beetles exhibit surprisingly diverse life history traits, including nonluminous adults with daytime activity periods, glowworm fireflies with flightless females, and lightning bugs that exchange species-specific flash signals. Fireflies also inhabit ecologically diverse habitats, including wetlands mentioned above. Their larvae, which can be aquatic, semiaquatic, or terrestrial, spend months to years feeding on snails, earthworms, and other soft-bodied prey. In contrast, firefly adults are typically short lived and do not feed. Some varieties are habitat and dietary specialists, whereas others are ecological generalists. Fireflies are economically important in many countries, including South Korea and Mexico, where they represent a growing tourist attraction . However, as is true for many invertebrates , fireflies have been largely neglected in global conservation.

Fireflies originally evolved the ability to light up as a way to ward off predators, but now they mostly use this ability to find mates. Not all species produce light, there are several that are day flying and apparently rely on the odors of their pheromones to find each other. Fireflies are filled with a nasty tasting chemical called lucibufagens, and after a predator gets a mouthful, it quickly learns to associate the firefly's glow with this bad taste. The flashing not only helps to attract a mate, but also warns predators to stay away. One species of firefly that cannot make its own lucibufagens acquires it by eating others that can. To lure victims, these fireflies mimic the flashing pattern of other species. When the unsuspecting make approaches to find a mate, instead it becomes a tasty treat to the tricky firefly. Each firefly species has its own signaling system. The males fly around at the right height, the right habitat and the right time of night to attract females. The females sitting on the ground or in vegetation watch for the males. When a female sees one making her specie's signal and doing it well, she flashes back with a species appropriate flash of her own. Then the two reciprocally signal as the male flies down to her. If all goes well they mate.

Fireflies glow is produced in photocyes or light cells located in the insect's abdomen. The light is the result of the chemical reaction termed bioluminescence which occurs when two substances, luciferin and luciferase react with one another when exposed to oxygen. The firefly regulates the flow of oxygen into its abdomen to turn its taillight on or off. Of note is that this cold living light is almost 100 percent efficient, losing only a fraction of its energy to heat. By comparison a standard incandescent light bulb is less than 10 percent efficient and an LED ranges between 40 and 50 percent.

Final thoughts

Insects like fireflies are predators that help to suppress pest populations or they are pollinators that help to produce the food we need. Their disappearance could create havoc with food cycles, especially for the birds and other animals that feed on them. A recent article in the journal Bioscience suggests a number of countermeasures for conservation. It is worth reading.

Locusts

—*A humanitarian crisis!*

AFRICA HAS RECENTLY been plagued with a devastation of the desert locust (*Schistocerca gregaria*). The swarms are the biggest in 25 years in Ethiopia and Somalia, and the worst Kenya has seen in 70 years. Further swarms have also been observed in Eritrea, Kjibouti, Tanzania, South Sudan and Uganda. Now swarms have reached Saudi Arabia and Pakistan. Locusts are the world's oldest pest and its most damaging destroying crops and threatening other economies. This is the most devastating outbreak in generations, it is threatening people's livelihoods. Even a small swarm can devour enough food for 35,000 people each day. Moreover, the regions affected are where approximately 19 million people already face food insecurity, and another 20 million are on the brink. The United Nations warns that the number of locusts could continue to skyrocket over the next few months and that the outbreak will lead to even more food shortages as planting season begins in the spring. It is a humanitarian crisis. Climate change may be at least partly responsible. Locusts thrive on warmer temperatures and heavy rainfall. With more extreme weather conditions, like recent cyclones, the locusts have migrated to larger areas. Making matters worse is that some countries like

Kenya and Ethiopia, lack the tools to combat the locusts. In Somalia, pest controllers cannot reach certain areas due to civil war. People in Ethiopia also continue to suffer from violence. The locusts are coming on the heels on what has been a very tumultuous year for East Africa. Last year the region swung between conditions that were either too hot or too dry or too wet.

An average swarm, which contains up to 40 million insects, can travel up to 150 km (93 miles) in a single day. Desert locusts live for about three months. Adults lay eggs that can hatch to form a new generation up to 20 times bigger than the previous one. The only method of controlling the spread of locusts is to kill them with insecticides at dusk when they land on bushes or trees or early in the morning.

Locusts or grasshoppers were once prevalent in the United States Midwest in the mid to late 1800s. The ultimate account of a locust invasion at that time is surely Laura Ingalls Wilder's description in her fourth book of the Little House on the Prairie series, *On the Banks of Plum Creek*. The incident occurred in 1875. "Laura's father regaling the family with the fertility of the country, the abundance of their crop, and the rosiness of their future—and then the locusts arrive. The insects' arrival is presaged by a strange foreboding: The light was queer. It was not like the changed light before a storm. The air did not press down as it did before a storm. Laura was frightened, she did not know why. She ran outdoors. A cloud was over the sun. It was not like any cloud she had ever seen before. It was a cloud of something like snowflakes, but they were larger than snowflakes, and thin and glittering. Light shone through each flickering particle. There was no wind. The grasses were still and the hot air did not stir, but the edge of the cloud came on across the sky faster than wind. The cloud was hailing grasshoppers. The cloud was grasshoppers. Their bodies hid the sun and made darkness. Their thin, large wings gleamed and glittered. The rasping whirring of their wings filled the whole air and they hit the ground and the house with the noise of a hailstorm."

The insect Laura Wilder was referring to is the Rocky Mountain Locust (*Melanoplus spretus*) that strangely became extinct. It was a migratory locust that in peak years spread over the Great Plains from Canada to Texas and periodically devastated the crops of homesteaders and farmers. Swarms of this insect often covered an area larger than Colorado. According to Jeffrey Lockwood in his book, *Locust*, the mystery began in the 19th century: instead of another invasion during the next drought cycle, the locust completely disappeared by the turn of the 20th century, without any apparent cause. He called it the quintessential ecological mystery of the North American continent. It was once the most abundant insect and it rivaled bison populations in both biomass and consumption of forage. Before the plains were settled, periodic swarms of migrating locusts were part of the natural rhythm of the grasslands, particularly during years of drought. That situation changed by the mid 1870s, however when farmers and ranchers occupied much of the Great Plains. A drought of several years duration triggered a massive outbreak of locusts that swept over an immense area, destroying much of the agricultural production and bringing famine to many settlers.

Final thoughts

One possible reason for the locust's disappearance is that the insect's eggs fail to hatch if they are deposited in earth disturbed by plowing or some other means. Cattle grazing may have also been involved.

Logic and rhetoric

—Two essential courses well worth studying.

SHAMEFULLY, THE LAST time I checked, few if any high schools or colleges were teaching courses dedicated to logic or rhetoric. Even though both would provide additional qualifications needed for college or employment in the medical or other fields. From my experience as a supervisor and someone involved in hiring, I believe that the best employees are curious, inquisitive and, most important, possessed with the ability to think logically. Logic is the pursuit of methods and principles used to distinguish correct from incorrect reasoning. It has been defined as the systematic study of the standards of good reasoning. With logic we learn how to acquire truths and to evaluate competing claims for truth. Using the methods and techniques of logic can distinguish between sound and faulty reasoning. The success of science rests on the ability to reason logically and every scientific method requires supporting logic.

Unfortunately formal logic is difficult to learn, it is fraught with syllogistic terms, symbols, quantification rules and inductive inferences. There are various forms of logic including informal, classical and modern.

Fortunately, modern or useful logic can be learned by watching DVDs from Great Courses or reading books including those about thinking fast and slow, inductive and deductive reasoning, scientific methods and problem solving.

Rhetoric, the other fundamental course, has a number of meanings; it is the study and practice of effective communication, the study of the effects of language on an audience, and the art of persuasion. It can be thought of as the capacity to produce appropriate and effective language in any situation and to present the results of logical thinking. Rhetoric, like logic, was once an essential part of the liberal arts curriculum. Like logic, rhetoric can be learned from on line courses or books. One I am now reading is Richard Toye's Rhetoric—a Very Short Introduction. Understanding rhetoric can be useful when applying for a job, interviewing, or writing a letter to gain admittance to a college or university.

Aristotle (384-322 BC), a Greek philosopher, is credited for discovering logic along with other contributions to science and philosophy. He wrote the first known treatises of logical theory and began teaching the first logic classes in history. The courses covered the standards any reasoning must follow if it is to be good reasoning. His theory of the syllogism has had an unparalleled influence on the history of Western thought. Syllogisms are structures of sentences each of which can be meaningfully called true or false. According to Aristotle, every such sentence must have the same structure: it must contain a subject and a predicate and must either affirm or deny the predicate of the subject.

Aristotle also taught rhetoric, his book on the subject is possibly the first work to be written on the subject and one of the seminal works of Western philosophy. He focused on the use of language as both a vehicle and a tool to shape persuasive argument, and emphasized the role of language in achieving precision and clarity of thought. Ancient philosophers regarded rhetoric as the crowning intellectual discipline—the synthesis of logical principles.

The ability to write a persuasive letter or to present a persuasive argument requires special skills, and many of us have not been taught the basic requirements. Fortunately, with regard to letters, there are instructions available on the internet that can guide you on the format, including such obvious matters as including your name and address and the name and address of the person you are writing to. Internet sites also tell you how important it is to be clear and concise and to lay out valid argument points. This is the point where logical thinking is required and where the internet is of little use. But , there is hope for all of us. According to a recent article in Science, we all have the inherent ability to think logically: it may be deep in our developmental roots. Even preverbal infants appear to draw inferences about objects and causes. The study indicated that one year olds can spontaneously reason using the process of elimination. Researchers found that the infant's pupils dilated more when watching movies that required rational deductions, a phenomenon that occurs in adults during deductive reasoning as well. Just imagine how much better we can be with additional education and training.

Mathematics and Numbers

—Shouldn't we be more numerate?

I CONSIDER MYSELF semi- innumerate (lacking many math skills) and that may be one of the main reasons I am so enamored with mathematics of all descriptions including statistics and calculus. The latter was defined by humorist Dave Barry as "the branch of mathematics that is so scary it causes everybody to stop studying mathematics." My library at home is filled with books of all descriptions about the history of mathematics, the birth of numbers, the joy of numbers, statistics, biostatistics, the science of measurement, calculus, precalculus and numerous others. I also have DVDs from Great Courses on the History of Mathematics. Despite this vast collection I remain confounded by the subject but more appreciative of it. I view it as a lingering challenge despite my advanced age and I encourage all people young and old to become more familiar with all of benefits derived from being more numerate. A good way to start is to read "Innumeracy", a book written by John Allen Paulos. Parents should also encourage their children to consider a career in mathematics, whether it be teaching, science, engineering, statistics, quality control, or annuities. There will always be opportunities for work in any of these fields.

I thought about math recently because of the train derailment in Washington state where there were a number of fatalities. When it occurred I wondered what calculations had been used to establish speed limits for trains negotiating curves. It was not difficult of find. Not surprisingly, a study at the University of Florida considered maximum velocity, railway track embankment, track curvature, centrifugal forces, friction and other factors to establish derailment speed. All of the various factors were determined using mathematical formulas and proved once again the ancient wisdom that "Mathematics is the way to understand the universe." Since the days of Galileo and Newton, math has nurtured science. Unfortunately, the United States is far behind other nations in mathematics and math teachers are in short supply. Studies have shown that American students score significantly lower than students worldwide in math achievement, ranking 25th among 34 countries. One reason may be that many of us agree with Dave Barry. Therefore, I have included a few examples where simple math can actually be surprisingly useful. They may even provide the incentive to learn more about the essentials.

PROBABILITY

Probability involves weighing up the chances or likelihood of something or another taking place. It can often become full of surprises. The birthday problem is a famously counter intuitive result. If 50 random people are in a room, there is a 97 per cent chance that at least two of them will have the same birthday. In this case it is easier to compute the probability that all the people have different birthdays, as follows:

$$\frac{365}{366} \times \frac{364}{366} \times \frac{363}{366} \ldots .. \frac{319}{366} \times \frac{318}{366} \times \frac{317}{366} = 0.03$$

The product of all the fractions is about 0.03. Thus the probability that no two people have the same birthday is about 3 per cent. Hence, the probability that at least two people *do* have the same birthday is about 97 per cent. A similar calculation using 23 people rather than 50 yields ½ , or 50 percent, as the probability that at least 2 of 23 people will have a common birthday.

Anyone who gambles should be acquainted with the probability theory and the notion of independence. Two events are said to be independent when the occurrence of one of them does not make the occurrence of the other more or less probable. If one flips a coin twice, each flip is independent of the other. If one rolls a pair of dice, the top face of one die is independent of the top face of the other. Calculating the probability of two events occurring is easy to do—-simply multiply their respective probabilities. As an example, the probability of obtaining two heads in a row in a coin flip is ¼—1/2 x 1/2. The probability of rolling a 2 (1,1) with a pair of dice is 1/36—-1/6 x 1/6, while the probability of rolling a 7 is 6/36 since there are six mutually exclusive ways [(1,6), (2,5)(3,4), (4,3), (5,2), and (6,1) in which the numbers on the faces can add up to 7, and each of these ways has the probability of 1/36 or 1/6 x 1/6. You can bet accordingly.

Finally, while writing this article I came across a new book with a quotation about mathematics worth repeating. It is as follows: "The laws of nature are written in the language of mathematics. Math is a way to describe reality and figure out how the world works, a universal language that has become the gold standard of truth. Hence those who are fluent in this new language will be on the cutting edge of progress."

Measles

—Back again!

MEASLES WAS DECLARED eliminated in 2000 in the United States, but for some strange reason there have been five recent outbreaks, for a total of more than 120 cases. There was also an occurrence in 2014 in nine counties in Ohio, where a total of 383 cases of measles were reported. These cases originated from two unvaccinated Amish men whom measles was incubating at the time of their return to the United States from the Philippines. In that same country early this year, measles has caused the death of seventy children. According to the Center of Disease Control and Prevention, measles is still common in many parts of the world, including some countries in Europe, Asia, the Pacific, and Africa. Worldwide an estimated 20 million people get measles and 146,000 people, mostly children, die from the disease each year. A recent measles epidemic in Madagascar has caused more than 900 deaths. One reason has been the unwillingness of parents to vaccinate their children on time. Those most at risk are infants from nine to eleven months old.

For anyone who has forgotten about the disease, measles is an infection of the respiratory and immune systems and the skin caused by the measles virus. Measles is exceptionally contagious with a substantial degree of morbidity and mortality. It can be transmitted by droplets from the infected person's nose or mouth. If you are in a room with someone infected with measles, you can inhale their virus when they cough, sneeze or even talk. Infected people can transmit measles virus starting four days before they develop a rash, and thus they may be contagious before they realize they have the disease and remain able to spread the virus for about four days after the rash appears. Ninety percent of people not immune against the virus but sharing living space with an infected person will catch it. Measles causes serious respiratory symptoms (asthma, breathing problems), fever and rash. In some cases, especially in babies and young children, the consequences can be severe. About one in four people in the U.S. who get measles will be hospitalized. One of every 1000 people with measles will develop brain swelling which could lead to brain damage. One in 20 children with measles develops pneumonia. Pregnant women with measles are at greater risk of having premature or low-birth rate babies. One or two out of 1000 people with measles will die, even with the best of care.

The symptoms usually develop 7-14 days after exposure to the virus. The initial symptoms usually include high fever (often greater than 104 F), Koplik spots (spots in the mouth that usually appear 2-3 days prior to the rash and last 3-5 days), weakness, loss of appetite, red eyes, runny nose, and sometimes cough. The telltale red dots appear on the skin, beginning on the face, then spreading down the body. Complications are usually more severe in adults who catch the virus, in malnourished and immune compromised individuals.

Measles remains a rare event in pregnancy in developing countries since most women of childbearing age acquired measles at a young age. However, in industrialized countries the age distribution of measles is changed by immunization, resulting in measles infection in young adults.

Measles was described by Muhannad ibn Zakariya ar-Razi (860-932), a Persian philosopher and physician in the 10th century A.D. as a disease more dreaded than smallpox. A Scottish physician, Francis Home, demonstrated in 1757 that measles was caused by an infectious agent present in the patient's blood. The virus that causes measles was isolated by Drs. John Enders and Thomas Peebles in Boston in 1954. Humans are the primary host, but non-human primates can also be a host, and measles is a threat to their conservation.

Even if your family does not travel internationally, you could still come into contact with measles anywhere in your community. Every year measles is brought into the United States by unvaccinated travelers (Americans or foreign visitors) who get measles while they are in other countries.

The best protection against measles is measles-mumps-rubella (MMR) vaccine, it provides long lasting protection against all strains of measles. Your child needs two doses of MMR vaccine for best protection. The first dose at 12 through 15 months of age and the second dose 4 through 6 years of age.

If your family is traveling overseas, the vaccine recommendations are somewhat different. If your baby is 6 through 11 months old, he or she should receive one dose of MMR vaccine before leaving. If your child is 12 months of age or older, he or she will need 2 doses of MMR vaccine (separated by at least 28 days) before departure.

MEDICAL QUACKERY

—Not just something from the past.

THE RECENT DEATH of William Helfand, a collector of memorabilia about bogus medical treatments in the United States dating back centuries reminded me of something I wrote many years ago. Mr. Helfand was an expert in the history of quackery and the methods for promoting it. The article I wrote was published in Medical Device and Diagnostic Industry magazine in 1983. Quack medical devices fraudulently promise to cure disease without evidence of safety or effectiveness. The wide spread availability of such products was instrumental in the enactment of the Federal Food, Drug and Cosmetic Act of 1938 and the Medical Device Amendments in 1976. Photographs of a number of them are available on the Pinterest website:

https://www.pinterest.com/sue57sandoval/quack-medical-devices-and-cures/?lp=true.

The first quack device may have been the first medical item patented under the Constitution. In 1796, the government granted Dr. Elisha Perkins a patent for metal rods called tractors, they consisted of a pair of small brass and iron instruments each about three inches long, flat on one side, rounded on the other, and tapered with a sharp point. One was gold in color, the other silver, the pair sold for about five guineas (around 25 dollars), extremely expensive in those days. Perkins claimed that his tractors used a mysterious force to unlock the secret door to health. The points of the tractors were drawn, first one, then the other across the afflicted surface of the body and allegedly drew the disease elements out. Considering the time period and the quality of medical practice that prevailed, it was not surprising that Perkin's treatment enjoyed amazing popularity.

The availability of quack devices continued through the 19th century, and could be deemed the "Age of Electrotherapeutics." In the belief that electricity could cure a number of ailments, it became common practice to deliberately send shock waves through the body. Much of this theory stemmed from the work of Luigi Galvani, whose research on isolated frog muscles showed they would react spasmodically when exposed to an electrical impulse. There were a wide variety of electrical quack devices, an example was the "Improved Magneto-Electric Machine for Nervous Diseases." Many similar devices could be found in doctor's offices and a jolt of electricity was often standard therapy.

In 1892, almost a century after Perkin's patent was granted, Dr. Hercules Sanche obtained one for another cure-all device. The "Oxydonor", according to the doctor, "caused the human organism to thirst for, and absorb oxygen, the vitalizer of blood." The oxydonor was a sealed metal cylinder containing a stick of carbon, and an uninsulated, flexible cord was attached at the other end. At the end of the cord was a small disc. To initiate the action the user merely placed the cylinder in a bowl of cold water and attached the disc to a wrist or ankle with an elastic band. The machine supposedly removed oxygen from the water and magically forced it through the "myriad pores" of the patient's body. It was indicated for all forms of disease and advertised in several of America's best monthly magazines.

Sanche was the forerunner for a host of other 20th century charlatans. The most notorious may have been Albert Abrams, founder of "Electronic Medicine" or "Radionics." Abrams claimed to have developed a diagnostic system that could determine the sex, age, race, disease, and even the religion of a patient he had never even seen. Even more miraculous was his invention of an apparatus called the "Oscilloclast." With this device Abrams claimed to be able to cure most of the ills of the flesh. Following the oscilloclast, his inventions became more sophisticated and he even published a book modestly entitled "The Electronic Reactions of Abrams", later shortened to the E.R.A. It disclosed the belief that the human body possesses rates of electronic vibrations of sickness and health. It followed that the type, severity, and location of any disease could be determined by measuring the perturbed vibratory rates. These rates, of course, could be measured by instruments designed by Abrams. For diagnosis, a drop of blood, preserved tissue, a photograph, or a sample of the patient's handwriting was needed. Abrams and his remarkable claims took the world by storm, and the E.R.A. became known throughout the civilized world. Drops of blood on blotters, accompanied by checks for ten dollars poured in by the hundreds to Abram's office in San Francisco.

Many worthless cure-all devices were still being sold in the 1950's and 60's. In fact the Food and Drug Administration and the AMA held a National Congress on Medical Quackery in 1961. Much of this changes with the enactment of the Medical Device Amendments of 1976, but we continue to witness the promotion of quack devices, albeit in a more limited fashion. The old time hokum still exists and will persist as long as vanity, dishonesty and incurable diseases endure.

MEDICAL WRITERS

FOR ANYONE WHO enjoys reading about medicine and science there are a number of writers, past and present, to search for. Two of my favorites are Lewis Thomas and Sherwin Nuland. Others more contemporary to consider are Natalie Angier, Atul Gawande and Gina Kolata. All of these authors express their thoughts with eloquence, simplicity and grace. Anyone with aspirations to become a medical writer should read some or all of their work. I used one of Lewis Thomas' books as a textbook in my medical writers course at Grace college.

LEWIS THOMAS

Lewis Thomas is perhaps the best known and most quoted medical essayist of the 20th century. He was a physician and biologist, widely known as the author of *The Lives of a Cell, The Medusa and the Snail, Late Night Thoughts on Listening to Mahler's Last Symphony* and *The Fragile Species*. All of these books are a treatise on biology as philosophy, and the work of a scrupulously observant, appreciative, self deprecating, and dryly funny mind. A series of his essays appeared in the *New England Journal of Medicine* beginning in the early 1970s with the modest title "Notes of a Biology Watcher." They were made available to the general public in two of his first books. In *The Fragile Species,* the last book written by Dr. Thomas, he contributed to an understanding of some of the great medical puzzles of our era, including AIDS, drug abuse, aging, and cancer. (He wrote about the latter from personal experience when he was diagnosed with a form of lymphoma in 1988.) Dr. Thomas disclosed that aging is not universal in nature, nor is it common. "Aging, real aging—the continuation of living throughout a long period of senescence—is a human invention and perhaps a recent one at that. It took us a long time and a reasonably working economy to recognize that healthy, intelligent old human beings are an asset to the evolution of human culture."

He described our phobia about germs as follows: "They will invade and replicate if given the chance, and some will get into our deepest tissues and set forth in the blood, but it is our response to their presence that makes the disease. Our arsenals for fighting off bacteria are so powerful, and involve so many different defense mechanisms, that we are in more danger from them than from the invaders. We live in the midst of explosive devices, we are mined."

Dr. Thomas' essays demonstrate the basic principles of a great writer's art: brevity, clarity, simplicity and humanity. Finding someone who writes as well and could be deemed his successor would appear unlikely. There is one, however, that is a potential candidate.

SHERWIN NULAND

Sherwin Nuland was a retired surgeon who taught bioethics and the history of medicine at Yale University. He is the author of *The Art of Aging The Mysteries Within, The Wisdom of the Body,* and the acclaimed best selling, National Book Award winning *How We Die*. Dr. Nuland practiced his craft in much the same manner as Lewis Thomas, but he wrote books instead of essays. He has been described as part poet, part philosopher, and part physician. In the *Wisdom of the Body*, he devotes one of the chapters to the fundamental unit of life, the cell. In the following excerpt he describes the cell's activities: "At any given instant, in any given cell, millions of molecular interactions are taking place. Were they not noiseless, the din emanating from the center of a cell's ceaseless tempest of surveillance, commands, and determined activity would be painful to the ear of some imaginary creature infinitely small enough to listen to it. The sound made by one of the body's organs would be intolerable, and as for a whole man—well he could be heard from the next county." The descriptive narrative, thought processes and provocative imagery provided in the writings of Drs. Nuland and Thomas are uncannily similar.

In an article written more than 35 years ago, Richard Asher, the author, suggested that writers take note of "The Elephant's Child," a Rudyard Kipling poem: "I keep six honest serving men (They taught me all I know): Their names are What and Why and When and How and Where and Who." These six serving men are used to attain most of our knowledge. Remembering who will read the work and then how to make it plain, simple, accurate, orderly, and complete are the keys to a clear writing style. It is a safe bet that both Drs. Thomas and Nuland read it and applied those principles in their classical works.

The Mediterranean Diet

—Worth adopting.

It is well known that a diet encouraging healthy eating and physical activity and that discourages alcohol consumption is associated with a reduced risk of breast, prostate, and colorectal cancer. Studies have also suggested that a calorie limited diet high in fresh fruits and vegetables, whole grains, tree nuts, legumes, olives and olive oil, few processed meats, low amounts of dairy products, low to moderate amounts of wine with meals, and low in animal protein, particularly red meat, can lower the risk of heart attacks and strokes, decrease chronic disease and extend life expectancy. This regimen, known for its cardiovascular disease fighting properties, has been deemed the Mediterranean diet. The Mayo Clinic calls it the "heart healthy diet." Nutrition has long been recognized as an important factor in healthy aging. One of the biggest problems in people as they age is loss of appetite that can lead to a number of health problems. Proper diet can play a vital role in keeping the elderly healthy and controlling healthcare costs as the number of adults aged 65 years or older grows in the coming decades.

There is also new research that found the Mediterranean diet enhances the good bacteria living in the gut. The study showed that Lactobacillus species, which are beneficial probiotic organisms, were significantly increased with a Mediterranean diet regimen. (We have about 2 billion good and bad bacteria living in our gut.) Such data are useful in further studies aimed at understanding the diet-health interactions in humans, including obesity, type 2 diabetes, cardiovascular disease and psychiatric disorders including Alzheimer's disease.

The Mediterranean diet should be the focus for all of us concerned about our health. Modern America is plagued by one of the highest obesity rates in the world and we fail to meet the life expectancy averages of almost every other developed nation. According to a recent article in the New York Times, the Mediterranean diet would result in huge improvements in human and environmental health and in rural economic stability. The author suggests resisting the amount of "landfood" meat, meat that is costly to produce. Moreover, feeding livestock requires large quantities of corn and soy which depletes the soil and the need for fertilizers that wash into surrounding watersheds, and degrading drinking water. Corn and soy production could be partially replaced by small seed cool weather grain like oats. This would go a long way toward locking in healthy soil and limiting erosion. Another of the author's ideas is to look underwater for our protein. This would be a diet where only a spare amount of animal protein is consumed and where that small amount comes from the sea. Recent evidence links two portions of seafood a week with lower blood pressure, lower LDL cholesterol and lower triglycerides.

A major consideration about the Mediterranean diet is

its anti-inflammatory property. This may be the reason for the effect on frailty in older adults. One example is wheat bran, considered a whole grain. The outer layer contains anti-inflammatory phytochemicals. Salmon and avocados part of the Mediterranean diet are abundant in omega-3-fatty acids that have anti-inflammatory properties. Newly pressed olive oil contains oleocanthal, a compound similar to drugs used for pain and discomfort. Another advantage of the diet is nutrition. The suggestion of two or more servings of fish provides Vitamin B12 and micronutrients including Vitamin A, β carotene, vitamin D, vitamin B6, vitamin E, folic acid, vitamin C and α tocopherol. These antioxidants prevent oxidative stress, a risk factor in frailty.

An added advantage of consuming the foods that are part of the Mediterranean diet is its inclusion of fiber rich foods. Consuming fruits and vegetables can reduce the risk of developing diabetes, arthritis and of course heart disease. People who eat more fiber simply have lower odds of dying. Fiber helps to feed the billions of bacteria in our guts and keeping them happy means our intestines and immune systems remain in good working order. Fiber may also help to reverse obesity. In a recent study published in the American Journal of Clinical Nutrition, investigators reviewed a number of clinical trials in which fiber was used to treat obesity. They found that fiber supplements helped obese people lose five pounds or more.

Unlike many popular diets, eating the Mediterranean way does not require eliminating fat from your diet. It doesn't limit your fat consumption at all, in fact. with the Mediterranean diet, you simply replace all of the bad fats you eat with good fats such as olive oil, nuts, seeds, and avocados. There are a number of cook books available with recipes that are delicious and easy to follow and well worth the investment.

Menstruation

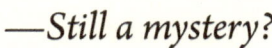

—*Still a mystery?*

THE WORLD IS full of strange customs, some the result of ignorance or lack of scientific progress. A recent example occurred in western Nepal, where women are banished from their homes every month during their menstrual period. They must sleep in huts for an entire week, and are not allowed to cook. They often sit by themselves and wait for family members to feed them. The women are considered polluted, even toxic, and an oppressive regime has evolved around this natural bodily function. The huts are as tiny as a closet with walls made of mud or rock. Earlier this year one woman died from smoke inhalation as she tried to keep warm by a small fire in the bitter Himalayan winter. Something similar occurs each year when at least one woman or girl dies in these huts, either from exposure, smoke inhalation or attacks by animals. The practice is called *chhaupadi*, from the Himalayan word that means someone who bears an impurity, and it has been going on for hundreds of years. Fortunately, the government and advocates for women are trying to end this bizarre practice. It may be difficult, however, because many people in western Nepal have been taught that any contact with menstruating women will bring bad luck. The isolation practice is thought to protect the purity of the community.

Although the strange custom in Nepal is uncommon, menstruation has long been a taboo and the topic often misunderstood. The belief that menstrual blood is toxic persisted well into the twentieth century. Menstruation has been used as an excuse to exclude women from different types of institutions. Earlier that century menstruating women were not allowed to enter churches around the world, wineries in Germany or opium laboratories in Vietnam.

Even today women may be reluctant to discuss the natural body function of menstruation. The process occurs once a month, when the lining of the uterus, acting on signals from estrogen and progesterone hormones, thickens with spongy, blood-filled nutrients. If the woman has had sex and an egg and a sperm join, this uterine lining (endometrium) will be used to sustain the developing embryo. If fertilization doesn't take place, the egg travels down the fallopian tube, through the uterus, past the cervix, and out the vagina. Approximately twelve days later, when the levels of estrogen and progesterone have dropped and the uterus has gotten the message that no pregnancy has occurred, the uterine lining—blood and mucus—simply flows out. In total, each period consists of four to six tablespoons of blood.

Menstruation is defined as the shedding of the superficial endometrium with associated bleeding that occurs in some species of placental animals when progesterone levels fall at the end of an infertile reproductive cycle. Menstruation is limited to higher female primates, some species of bats, and the elephant shrew. In humans, it occurs once a month for about forty years.

According to Karen Houppert in her book, *The Curse,* "before the age of scientific knowledge, myths provided explanations for why young girls and women bled from their vaginas on a monthly basis. " Though these stories portrayed menstruation as a powerful process, ancient people also perceived it as a marker of women's inferiority and that vaginal bleeding was the sacred remains of an unborn child. Thus it was condemned as evil and dangerous.

Even today, advertisers and manufacturers tiptoe around the actual words, vaginal bleeding or menstruation. Commercial menstrual products are commonly referred to as feminine "protection"; but begs the question, protection against what?

A journey through the coded history of sanitary protection makes for a fascinating crash course in American sexuality—and its repression. Shame and secrecy are the primary message. Disposable pads owe their origin to nurses who first thought of holding the flow of blood with available wood pulp bandages used in hospitals. These bandages were made by Kimberly Clark for American soldiers. Nurses in France during World War 1, used them as menstrual pads. because of their absorbency, and cost. Manufacturers of bandages borrowed the idea and produced pads made from handy products inexpensive enough to be disposable. The first of the disposable pads were in the form of cotton, wool, or similar fibrous material covered with an absorbent liner. The liner ends were extended front and back to fit through loops in a special girdle or belt worn beneath undergarments. Kotex, first called Cellucotton and Cellu-naps, were marketed around 1920. They weren't accepted until about 1926 when Montgomery Ward advertised the product in its catalogue. Although available, disposable pads were often too expensive for many women to afford. It took several years for such pads to become commonplace.

FINAL THOUGHTS

Elissa Stein in her book *Flow*, comments how menstruation has continued to remain a complex event, and that strangely, doctors and scientists have yet to fully understand it or what exactly it does.

Adopting Mindfulness Training

—Why not?

I HAVE HAD a number of disappointments in my career but perhaps one of the worst one came with my inability to convince manufacturers of medical devices to adopt mindfulness training for all employees. The major purpose of such training is to reduce human errors that occur in the work place. My effort consisted of personal visits for presentations at three local firms, discussions during my compliance courses at Grace college and writing three articles on the subject, the last of which appeared in Quality Progress magazine. That article was optimistically entitled "The Next Phase in Quality's Evolution" and it described current activities, history, definition, purpose and goals, testimonials, training, and what I deemed the future benefits. Quality Progress is a peer reviewed publication of the American Society for Quality (ASQ). The society has more than 90,000 members, most anyone with quality responsibilities joins ASQ. I eagerly waited for feedback—- almost none was forthcoming. In fact, just three individuals sent comments and one of those justifiably was about the quality of the photograph accompanying the article.

Needless to say, my expectations greatly exceeded reality and I should just accept public indifference. However, I am stubborn enough to persist in supporting a training method with more than just the purpose of reducing errors. Mindfulness, in my opinion, can prove beneficial to everyone and more and more people are adopting the practice. Let me first define the term: mindfulness is awareness, cultivated by paying attention in a sustained and particular way; on purpose, in the present moment and nonjudgmentally. It is one of many forms of meditation. A common way of describing mindfulness is to think of it as the general receptivity and full engagement with the present moment. Many companies have mindfulness training programs in the USA, including among others, Google, Target, Chase bank and the Mayo Clinic. There has been a surge of interest during the past decade driven by what has been called distraction addiction, where on a typical day individuals may send or receive 110 emails, check phones up to thirty of more times and visit Facebook and Twitter frequently. Americans, on average, idly spend 60 hours on line. People tend to use half of their time not thinking about the task at hand, even when they are told explicitly to pay attention. Multitasking (trying to do two or more things at once) is another reason for making mistakes, it disrupts the kind of sustained thought required for problem solving and creativity. In fact, research has shown that pigeons are better at multitasking than humans. Moreover, we know that the average attention span has fallen to about eight seconds, down from twelve seconds in the year 2000. According to a recent study, we have a shorter attention span than goldfish.

I firmly believe that mindfulness meditation can reduce errors: it reminds us of what we should be doing and to pay full attention. We see things as they really are. Fortunately, there are other benefits as well. Controlled studies suggest that mindfulness meditation can effectively reduce symptoms in people with chronic pain, recurrent depression, anxiety disorders, substance abuse, binge eating, and many other health conditions. It can even change the brain's grey matter in ways that control conscious control over emotion. Increasingly it is evident that mindfulness can also benefit cardiovascular health by virtue of its effect on reducing stress, blood pressure and obesity. One small study even indicates that meditation can help slow the progression of Alzheimer's disease. Many psychologists believe that people who meditate regularly enjoy life more and appear to be happier.

Anyone interested can employ the following steps to practice mindfulness:

1. Sit upright in a chair in a stable position, with hands resting on the thighs.
2. Lower or close your eyes, whichever is more comfortable. (Closing is preferable.)
3. Attend to your breath, following its movement throughout the body.
4. Notice the sensations around your belly as air flows into and out of your nose and mouth.
5. Select one area of your body affected by your breathing and focus your attention there. Control your focus, not the breathing itself.

6. When you notice your mind wandering—and it will—bring your attention back to your breathing.

7. After five or ten minutes, switch from focusing to monitoring. Think of your mind as a vast open sky and your thoughts, feelings and sensations as passing clouds.

8. Feel your whole body move with your breath. Be receptive to your sensations, noticing what arises in the moment. Be attentive to the changing quality of experience—sounds, aromas, breezes, and thoughts.

After about five minutes, lift your gaze and open your eyes.* Daily practice of ten minutes or more is recommended.

*Reference: Scott Rogers. Mindfulness Research and Practice Initiative. in Amishi P. Pha, "Being in the Now, Scientific American Mind, April/May 2013.

Mosquitoes

―――∽∽∽―――

Pundits are predicting that global warming will result in unimaginable problems, including the possibility that tropical diseases will become much more prevalent in the United States. Rainforests in Africa or in the Amazon region could also disappear in this century and create new breeding sites for certain disease related insects. If either occurs, and both are likely, mosquitoes will no doubt be the vectors. Mosquitoes are our most persistent and deadly enemy. They have killed great leaders, decimated armies, and decided the fates of nations. Alexander the Great, for example, was likely killed by malaria in 323 B.C. Mosquitoes are responsible for a host of devastating, difficult- to- treat diseases, including dengue, elephantiasis, and malaria, diseases which affect more than 10 percent of the world's population. Mosquitoes are unique in terms of their numbers, varieties, life-cycles, geographic distribution, appetites, and habits. They are also attracted to some people more than others. Fortunately mosquitoes can be repelled by a number of chemicals and their repellant properties can be predicted from the chemical structure. One bite is all it takes to be infected with West Nile virus or other mosquito-borne diseases.

Diseases

Mosquito-borne diseases include malaria; Japanese, California, Eastern Equine, Western Equine, St Louis and Venezuelan encephalitis; dengue and dengue hemorrhagic fever; Rift valley fever; yellow fever; elephantiasis; and Chikungunya fever. The pathogens that cause these diseases are transmitted by injection of saliva into susceptible hosts by female mosquitoes needing protein from a blood meal to develop their eggs.

Mosquitoes are responsible for diseases in more than 700 million persons each year. Malaria alone kills 3 million persons each year, including one child every 30 seconds. In the United States, arboviruses (arthropod- borne viruses) transmitted by mosquitoes continue to cause sporadic outbreaks of eastern equine encephalitis, western equine encephalitis, St. Louis equine encephalitis, and La Crosse encephalitis. Eastern equine encephalitis is the most lethal of the mosquito –borne encephalitides. The infection first became evident in the suburbs of Boston in 1938, but the virus had been isolated from horses five years earlier. In 1999, West Nile virus was discovered for the first time in the New York city area where 62 persons were infected. The virus has now been detected in 27 states, and is expected to spread unabated across the United States.

Repellents

In many circumstances, however, applying repellents to the skin is an advisable way to protect against insect bites. The best known and most widely used chemical insect repellent is DEET. It was developed in 1953 and remains the gold standard based on its excellent human use safety record. The mechanism of action, however, has remained a mystery. Just recently, several researchers have discovered that DEET blocks odorant receptors in the nervous system of insects, masking odors that would ordinarily attract mosquitoes. Thus, mosquitoes smell and deliberately avoid DEET.

DEET, however, may be washed off by perspiration or rain, and its efficacy decreases dramatically with rising outdoor temperatures. DEET is also a plasticizer, capable of dissolving watch crystals, frames of glasses, and some synthetic fabrics.

Attractions

Body temperature, carbon dioxide in the breath, and certain skin chemicals, such as lactic acid, all help mosquitoes orient and find their next blood meal. This means that exercise which boosts the levels of all three signals makes us more appealing than just sitting around. Evidence also suggests that mosquitoes are more attracted to black or red clothing. White is the best choice. It is less certain whether perfume or other body products attract mosquitoes, but it may be wise to avoid products with floral scents. New research also indicates that pregnant women are twice as attractive to mosquitoes as those who are non-pregnant. Pregnant women in an advanced stage exhale 21 percent greater volume and mosquitoes are attracted to the moisture and carbon dioxide in exhaled breath. Moreover, the abdomens of pregnant women are 0.7 C hotter than the abdomens of women who are non-pregnant which suggests that pregnant women release more volatile substances from their skin, allowing mosquitoes to detect them more easily. There is recent research indicating that children with malaria are even more attractive to mosquitoes because they emit specific odors from their skin and thus invite further bites and risks of infection. Identifying these human derived compounds could provide an opportunity to use them for chemical lures to trap mosquitoes. Until then, the best bet for preventing bites include wearing long sleeves and pants, using repellents, and staying away from shady areas and vegetation.

MYSTERIES

The key to combating the scourge brought on by mosquitoes is to better understand them. After more than a century of research there are a number of mysteries. No one has explained why mosquitoes use blood in their reproductive process. Nor do we know how mosquitoes distinguish between hosts. Moreover, we do not understand the basics, such as how a mosquito develops the sucking force necessary to pull blood through a feeding tube with a diameter so small that resulting friction should make it impossible. In addition, it is still a mystery as to why certain species of mosquitoes support the development of one type of pathogen and not another.

Natalie Angier

—*A writer worth reading.*

THOSE OF US interested in science and medical writing should be aware of a contemporary author, Natalie Angier. I just finished reading for the second time, her book published in 1995 entitled *The Beauty of the Beastly*, a compilation of articles that originally appeared in the New York Times. She is also the author of *Natural Obsessions, Woman: An Intimate Geography*, and her latest, *The Canon*. I believe all of them are available in paperback. *The Canon* delves into physics, chemistry, biology, geology and astronomy. It is a primer on science and full of wit and wordplay. Angier writes, "science is not a body of facts but a state of mind, noting that researchers typically recognize the provisional nature of discoveries, revel in skepticism and are spurred by uncertainty." The *Beauty of the Beastly* contains new views on the nature of life. The book includes essays about dolphins, orchids, oxytocin the cuddling hormone, parenting, DNA, longevity, scorpions, roaches, pit vipers, hyenas, cheetahs, DNA and famous scientists, among others. Few writers have ever so vividly described each of those subjects. The New York Times Book Review writer noted that Natalie Angier is one who is constitutionally incapable of writing a boring sentence. For example , in an essay published in *The New York Times*, Angier described metastatic cancer thusly: "They are barbarians, the colonist cells, co-opting all nutrients in their adopted organ and starving their normal neighbors of air, sugar and salts, and blocking traffic and clogging conduits, and finally, when their greed exceeds their easy grab, tearing open surrounding cells and feasting like cannibals on the meat of their fellows." In her essay on the anatomy of joy, Natalie wrote that happiness is regarded as healthy only because it spares us the enfeebling impact of anxiety or inspires us to cultivate such worthy habits as eating vegetables, avoiding liquor or cigarettes, and sleeping eight hours a night. When commenting on the effectiveness of insecticides to kill roaches, she noted that in recent times many city dwellers have been able to stride into their kitchens at night with the newfound confidence that they can flick on the lights, take a glass from the cupboard, even grab a few cookies from a box on the counter—-all without the odious sight of dozens of greasy brown cockroaches scattering for cover.

Like Lewis Thomas, who I have written about earlier, Ms Angier essays demonstrate the basic principles of a great writer's art: brevity, clarity, simplicity, and humanity. Finding any current author who writes as well would appear unlikely. For anyone concerned about time wasted napping or goofing off, Angier wrote that those of us who feel the urge to take it easy but remain hardened to the work ethic might do well to consider that laziness is perfectly natural, perfectly sensible, and is shared by nearly every other species on the planet.

Natalie Angier's most recent article described what it means to be friends. She said that researchers have long known that people choose friends who are much alike themselves in a wide array of characteristics: of similar age, race, religion, socioeconomic status, educational level, political leaning, pulchritude rating, even handgrip strength. The impulse toward bonding with others who are the least other possible, is found among traditional hunter-gatherer groups and advanced capitalist societies alike. Even the brains of close friends respond in remarkably similar ways.

I have always worried about my fear of anything to do with numbers, until I read Angier's essay on animal's ability to count. In it, she explained that despite the prevalence of math phobia, people too are born with a strong innate number sense, and numerosity is deeply embedded in many aspects of our mind and culture. Even the words for small quantities are strikingly similar across virtually every language studied, and the words are among the most stable, unchanging utterances in any lexicon.

No greater praise for Angier's work, could be topped by the words by Richard Dawkins, found on the cover of *The Canon*, he wrote "Every sentence sparkles with wit and charm. But there's passion in there too, and it all adds up to an intoxicating cocktail of fine science writing." I can add that her work is worth reading over and over.

Many years ago Ms. Angier thanked the New York Times for allowing her to write about oddball subjects that few other newspapers would touch. I assume that readers of her work like myself would prefer to thank her. Since then Angier has won the Pulitzer Prize, an American Association for the Advancement of Science journalism award and many other honors. She is truly one of the great science writers of our time.

The day the dinosaurs died.

A number of exciting scientific achievements were reported last month, including the first photograph of a black hole, unearthing of a fossil belonging to a previously unknown cousin to *Homo sapiens* who lived 50,000 or so years ago, and the discovery of a new fossil site in North Dakota that marked the extinction of dinosaurs 66 million years ago. The latter event received the most press and rightly so. It may happen again, and if so, it would be mean the end of civilization as we know it. Just imagine what it would be like to watch a small glowing moon- like object traveling 45,000 miles per hour approaching the earth and growing larger and larger. When it strikes it will blast a hole in the atmosphere and generate a supersonic shock wave. The aftermath would be firestorms that incinerate the landscape for miles and miles. Even creatures thousands of miles away would be doomed, if not by fire and brimstone, then by mega-earthquakes and waves of unimaginable size. This may have happened when a gigantic asteroid struck a shallow sea off the coast of Mexico where the Yucatan peninsula is today. At that very moment the Cretaceous period ended and the Paleogene period began.

The event was described in a fascinating article in the New Yorker magazine on April 8. According to the author, computer models predicted that, "within two minutes of slamming into the Earth, the asteroid which was at least six miles wide, had gorged a crater about eighteen miles deep and lofted twenty five trillion metric tons of debris into the atmosphere. The energy released was more than that of a billion Hiroshima bombs, but the blast looked nothing like a nuclear explosion, instead the initial blowout formed a "rooster tail", a gigantic jet of molten material, which exited the atmosphere, some of it fanning over North America. Much of the material was several times hotter than the surface of the sun and it set fire to everything within a thousand miles. In addition something truly strange occurred. An inverted cone of liquefied, superheated rock rose, spread outward as countless red-hot blobs of glass, called tektites, blanketed the Western Hemisphere. Within hours or perhaps minutes of the titanic collision, sea creatures were swept inland by tsunamis and earthquakes, tossed together and deposited with a diverse array of landlocked life including trees, flowers and vanished types of freshwater fish. The tektites raining into water, clogged the gills of fish, which were then killed by surges of water." These contents formed the North Dakota deposit that paleontologist Robert DePalma uncovered and were described in the New Yorker article. As evidence, the gills of the fish were isotopically dated to 65.8 million years ago. As the earth rotated, the airborne material converged on the opposite side of the planet where it fell and set fire to the entire Indian subcontinent. About seventy five percent of all species went extinct, more than 99.9999 percent of all living organisms on earth died.

Much of what DePalma found confirmed work done in 1980, when Luis Alvarez and his son Walter, a geologist made public their theory that the dinosaurs (except birds) were killed by a massive asteroid strike. In the early 1980s, the discovery of a clay layer rich in iridium, an element found in asteroids, at the end of rock record of the Cretaceous at sites around the world led researchers to link an asteroid to the End Cretaceous mass extinction. A wealth of other evidence has persuaded most researchers that the impact played some role in extinctions. But no one other that DePalma has found direct evidence of its lethal effects. DePalma and his colleagues say the killing is captured in forensic detail in the North Dakota deposit, which they say formed in just a few hours, beginning perhaps 13 minutes after impact.

DePalma and others have been working on their find since 2012 and he also recruited Walter Alvarez to the team. Alvarez provided the meticulous archeological approach required for such a massive undertaking. It should be mentioned, however, that other scientists question DePalma's interpretations. According to Blair Schoene, a geologist at Princeton University, the site does not definitively prove the impact event alone triggered the extinction. He and some others think the environmental turmoil caused by the volcanic activity in what is now central India may have taken a toll even before the impact. And until a few years ago, some researchers had suspected the last dinosaurs vanished thousands of years before the catastrophe. Several more papers about DePalma's discovery are now in preparation and they will likely more fully describe the significance of the fossil findings. It may be the best evidence yet that at least some dinosaurs may have been alive to witness the asteroid impact or that dinosaurs truly died on that very day.

NEW DISCOVERIES IN HUMAN ANATOMY

ANDREAS VESALIUS' BOOK *De Humani Corporis Fabrica,* whose title is best translated as *On the Workings of the Human Body,* paved the way for modern scientific medicine. The magnificently illustrated book published in 1543, combined science, technology, and culture in a way few other books have ever done. It sparked a scientific revolution that altered history and is arguably the best known book produced in the history of western medicine. Vesalius was as interested in the functions of the human body as he was about anatomy and the book gave the world its first accurate knowledge of anatomy and a method by which it could be studied. Just a few years later, in 1550, Ambroise Pare wrote a similar book, entitled Brief Collection of Anatomy. Like Vesalius, Pare's book provided a full description of the human body. Modern medicine owes much to both writers as they debunked previous theories about anatomy prevalent since the time of Galen beginning in the first century. (Galenism is a system of medicine consisting of 84 technical treatises and the theory of four bodily humors, blood, phlegm, black and yellow bile.) Galenism was based on animal rather than human anatomy but his theories held sway for 14 centuries. In addition to Vesalius and Pare, great artists like Da Vinci,

Michaelangelo, Titian and Raphael were known for their work depicting the functioning human body. Each of them contributed to current knowledge about anatomy.

One would believe that after almost five hundred years of study, few secrets remain about the composition and structure of the human body. Not so, in 2013, a team of Belgian scientists discovered a new ligament located just outside the knee. To find and characterize this component, orthopedic surgeons Dr. Steven Claes and Dr. Johann Bellemans and their colleagues gathered 41 knee joints from human cadavers and began minutely dissecting them. Their discovery confirmed work done as far back as 1879, when Paul Segond, a French surgeon first speculated that in addition to the four obvious structural knee ligaments then known, the anterior cruciate, medial collateral, posterior cruciate, and lateral collateral, other ligaments must exist in the knee or it would not be stable. He wrote that during dissections he had noticed a "pearly, resistant fibrous band, originating at the outside, front portion of the thighbone, and continuing to the shinbone, which he felt stabilized the outer part of the knee. He did not, however, give this pearly band a name and somehow, in the decades that followed, its existence was forgotten or ignored. It took almost 150 years for resolution.

Last year, there were two other surprising discoveries. The first was a new ligament on the lateral side of the ankle. According to the guidelines of human anatomy, the ligaments in the ankle are grouped by two ligament complexes: the lateral collateral ligament in the side of the joint and formed by three independent ligaments and the medial or deltoid collateral ligament. In a new scientific study a research team defined a surprising anatomic structure in the ankle, the lateral fibulotalocalcaneal ligament complex. This was possible thanks to the analysis of fibers that link two of the lateral collateral ligament compounds. The discovery changes the understanding of the ankle joint and could explain why many ankle injuries produce chronic pain.

For the second, researchers in Sweden identified four types of neurons in the peripheral auditory system of the ear, three of which are new to science. The analysis of these cells can lead to new therapies for various kinds of hearing disorders, such as tinnitus and age-related hearing loss. It also opens the way for developing genetic tools than can be used for new treatments for other disorders and that influence the function of individual nerve cells.

This year, a network of very fine blood vessels that connects bone marrow directly with the blood supply of the periosteum (connective tissue covering bone) that was previously overlooked was discovered by researchers in Germany. Although bones are very hard substances they also have a dense network of blood vessels inside them where the bone marrow is located as well as on the outside that is covered by the periosteum. (This is the reason why bone fractures often cause serious bleeding.) These researchers have found previously unknown blood vessels in the bones of mice that that travel across the entire length of cortical bone. This newly discovered system of vessels is used by the immune cells in bone marrow to reach the blood stream and key to allowing immune cells to reach the site of inflammation quickly.

There may, of course, be other surprises in the future. Discovery techniques and instrumentation continue to improve as does the means of communicating scientific information.

Thanksgiving

—*What to be thankful for.*

When or if someone next Thanksgiving holiday asks you what you are thankful for, I have a suggestion. Why not describe all the things your body does to keep you healthy and alive, the automatic processes you have no control over. You could start with the immune system which consists of two different responses to anything that enters the body that doesn't belong, *innate* and *adaptive*. The innate immune system is responsible for the first wave of defense when confronted with an invader, whether it is bacteria, virus or venom. The adaptive system remembers previous attackers, and allows the body to mount a better response the next time those invaders attempt to reenter the body. As soon as invaders are detected the innate system takes action. Mast cells, packed with histamine and heparin and macrophages engulf anything that does not belong there. Inflammation is triggered—it is a carefully regulated response to kill some types of bacteria and viruses without damaging the body. Macrophages eat bacteria, viruses, and other foreign particles. If the macrophages do not destroy all of the invaders, they release compounds which attract neutrophils, a type of white blood cell. As the battle continues between the innate system and the unwelcome visitors, the adaptive system kicks in. It includes dendritic cells and T cells. All T cells contain have specialized receptors and only those whose receptors match what the dendritic cells are presented will be activated. Some will remain as memory T cells, while others become killer T cells to help macrophages and neutrophils. A third subset activate B cells, which are the body's antibody factories. With the combined attack of the antibodies, T cells, macrophages, and neutrophils, the invaders should be defeated. Just imagine trying to orchestrate that cascade under your direction. Moreover, you should recall that before the birth of modern medicine, it was the patient's immune system that cured him or her, or that didn't.

When you finish profusely thanking the immune system you can extol the virtues of your liver. According to Natalie Angier from the New York Times, its to-do list is second only to the brain and numbers well over 300 items. This includes systematically reworking the food we eat into usable building blocks for our cells; neutralizing harmful potentially harmful substances we incidentally ingest; generating a vast store house of hormones, enzymes, clotting factors and immune molecules; controlling blood chemistry, and so on. There is no machine available to replace all of the liver's diverse functions. The liver is our largest internal organ, weighing three and half pounds and measuring six inches long. It is always flush with blood, holding about 13 percent of the body's supply at any given time. Finally, it is the liver's responsibility to keep track of the body's moment- to- moment energy demands, releasing glucose as needed from its stash of stored glycogen, along with any vitamins, minerals, lipids, amino acids or other micronutrients that might be required.

If you still have an audience at the dinner table you can then praise your pancreas. It, indeed, is a miraculous organ and performs a number of vital functions without our instructions or knowledge. The pancreas is composed of two main elements, exocrine and endocrine tissues. The exocrine tissue is organized into a large number of sac-like structures lined with cells that secrete various enzymes important to the digestive process. As these pancreatic juices are made, they flow into the main pancreatic duct connecting the pancreas to the liver and gall bladder. Scattered throughout the exocrine tissue are small, isolated pockets of endocrine tissue, these pockets are known as the islets of Langerhans. Islets may be composed of several types of cells, predominantly alpha and beta cells. Granules in the beta cells produce insulin, while those in the alpha cells provide glucagon. Insulin helps control carbohydrate metabolism, while glucagon counters the action of insulin. Strangely, the pancreas can churn out huge quantities of enzymes to rapidly reduce our fast food diets into particles of amino acids, carbohydrates and fats, miraculously without digesting its own tissue in the process.

Any unlikely remaining individual could then be lectured about the central nervous system, blood, hormones, respiration, and circulation. With regard to the former, one author summarized the central nervous system as essentially a number of masses of nerve cells connected to each other by a complex set of fibers. The function of a nerve cell is to interpret the impulse brought by the nerve fiber, and to initiate new impulses to be sent out over other nerve fibers. He indicated that his summary contains everything that matters; none of which are under our immediate supervision. We indeed have a lot to be thankful for.

The surprising octopus

—One of nature's marvels!

Recent news that the population of octopuses in the world's oceans have increased significantly for the past fifty years sparked my curiosity about these unusual creatures. For example I didn't know that they possess unusual motor skills and intelligence. Further research indicated that a number of books have been written about them and other cephalopods. I learned that octopuses (not octopi) and humans have a common ancestor, possibly a slug like organism, that lived more than 550 million years ago, and that the octopus has evolved a camera eye closely similar to ours, but radically more elegant in design. The ancient Greeks christened the animal *oktopus,* which means, unimaginatively and inaccurately, "eight foot." The correct label for the appendages is actually "arms." This and other brief facts, e.g., why octopuses don't tie themselves in knots, will be included in this article and intended for curious individuals with an interest in surgery, evolution and zoology.

Octopuses belong to the class Cephalopods, an ancient group of invertebrates (animals that lack a spinal column) that appeared in the late Cambrian period several million years before the first primitive fish began swimming in the ocean. All cephalopods inhabit marine environments, none are found in fresh water. All are strictly carnivorous, and most hunt for prey. They generally grow quickly and have a short life span. As a consequence they do not have a chance to pass on what they learn to the next generation.

Cephalopods have gangly limbs encircling their mouths, the octopus possesses cone-shaped limbs (arms) with rows of suckers or suction cups. An octopus, like other cephalopods, has three hearts, two pump blood to the gills, and one central heart pumps oxygenated blood to the body. Like other cephalopods, octopus blood is blue because it binds oxygen using a blue, copper containing protein called hemocyanin. (Human blood is red because the oxygen binding protein hemoglobin contains iron.)

The hundreds of highly sensitive suction cups and the prehensile (capable of grasping or seizing) arms of an octopus are as effective in holding and manipulating objects as the human hand. The cups will stick to just about everything, with one important exception. They will not grab onto the octopus itself, otherwise the animal would find itself entangled and helpless. Researchers observed the behavior of amputated octopus arms, which remain very active for up to an hour after separation. Those observations showed that the arms never grabbed octopus skin, though they would attach to a skinned octopus arm. Apparently a chemical produced by octopus skin temporarily prevents their suckers from adhering. According to the report, the results show for the first time that the skin of the octopus prevents octopus arms from attaching to each other or to themselves in a reflexive manner. The octopus, surprisingly, can vary the stiffness of its arms, temporarily transforming the flexible limbs into stiffened segments to allow the octopus to move and interact with objects.

Cephalopods may have brain cells similar to humans and some species have been used to study how our brains work. If similarities are found, it would significantly alter the perspective on the emergence of life elsewhere in the universe.

Octopuses are considered by many to be the most intelligent of invertebrates, they have highest brain-to-body mass ratios. In addition, they have the manipulative ability to get into fishermen's crab traps, eat the crabs, and escape. They can solve simple mazes, open childproof bottles, build dens and can dismantle the aquariums they are kept in and can grow a new arm when one is bitten off. Once the arm is regrown it is basically as good as new. Octopuses have been known to throw rocks and smash aquarium glass, and on occasion to cause short circuits by crawling out of the tank and shooting a jet of water at an overhead lamp. Octopus dexterity is a marvel.

Octopuses are among the few animals in the world that can change color, matching it with their surroundings, rendering them nearly invisible, or alternatively give themselves a pattern that makes them stand out. Many thousands of color-changing cells called chromatophores just below the surface of the skin are responsible for these remarkable transformations. The most obvious reason to change color is to hide from predators. An octopus can change not only its coloring, but also the texture of its skin to match rocks, corals and other items nearby.

Throughout the years scientists have gained greater insight about the intelligence of the octopus. Recently, the octopus has become a new scientific model for researchers looking to study tissue regeneration. Findings could enhance progress in the Holy Grail of modern biology and create a storehouse of repair parts for the body. Octopuses are utterly different from all other animals, even ones they are related to. Moreover, they have somehow gained seemingly extraterrestrial capabilities and represent the pinnacle of an evolutionary track alternate to man and there is still much to learn from these marvelous creatures.

Otto Warburg

—*Another medical genius.*

I have previously written about a number of outstanding doctors and research scientists, including James Parkinson, David Nachmansohn, Emory Rovenstine, John Sulston and others, but Otto Warburg may be the most important of them all. His contributions to cancer research have become more widely accepted and continue to grow even though he died almost fifty years ago. He may have been the greatest biochemist of the 20th century. New insights into how cancer cells grow stem from Dr. Warburg's early work and have opened novel possibilities for treatment. In the 1920s, Otto Warburg and his colleagues discovered that cancer cells consume larger amounts of glucose (blood sugar) than normal cells. Glucose is the body's preferred energy source. (Table sugar or sucrose contains both fructose and glucose.) To generate energy from glucose, cells use one of two pathways, one of them takes place in the energy producing structures inside cells (mitochondria), and the second method is called fermentation. Normal cells use the path in the mitochondria, but about eighty percent of cancer cells seem to have revamped their metabolism to generate energy with fermentation. This phenomenon is known as the Warburg effect.

An article published in Science several years ago described the difference between normal and cancer cells. Normal cells take up nutrients from their environment when stimulated by growth factors. Cancer cells, on the other hand, acquire genetic mutations that alter the metabolic processes. These mutations result in the ability to utilize glucose to meet the demands for cell growth and reproduction just like the process of fermentation described by Louis Pasteur much earlier. Warburg found that unlike most normal tissues, cancer cells tend to "ferment" glucose into lactate and obtain the required energy needed for survival. He also proposed that cancer cells develop a defect in mitochondria that leads to impaired aerobic (the need for oxygen) metabolism. It should be understood that mitochondria are the power generators residing in our living cells and that by using oxygen to burn food, they produce all the energy we need to live and support growth. There are usually hundreds or thousands of them in a single cell, and mitochondria are present in every cell of the body with the exception of red blood cells.

OTTO WARBURG (1883-1970)

Otto Warburg was born on October 8, 1883, in Germany. His father was a famous physicist. Emil Warburg. Otto studied chemistry and was awarded a Doctor of Chemistry degree in 1906, he then studied medicine and obtained the degree Doctor of Medicine in 1911. He served in the Prussian Guard in World War 1 with distinction. (When Warburg enlisted in the army, Albert Einstein sent him a letter urging him to come home for the sake of science.) Warburg's early research was in botany, studying the assimilation of carbon dioxide in plants. He later went on to study the metabolism of tumors, and the chemical constituent of oxygen transfer during fermentation. For his discovery of the nature and mode of action of the respiratory enzyme, he won the Nobel Prize in 1931. (He was considered for the award on at least two other occasions.) This discovery opened up new ways in the fields of cellular metabolism and respiration. He showed for the first time, that cancerous cells can live and develop, even in the absence of oxygen. This research at the time was hailed as a major breakthrough in understanding cancer, but was largely neglected through the years and unpublished in textbooks. It should be noted that Warburg's work was conducted in Germany, despite the fact that Warburg was Jewish, an indication of how important cancer research was in that country. Cancer was more prevalent in Germany than in almost any other nation. While he was in the laboratory other members of his faith were being persecuted or awaiting death in Nazi concentration camps.

Through the years it became more obvious that cancer cells were hungry for glucose. His discovery, later named the Warburg effect, has been found to occur in up to 80 percent of cancers. It is so fundamental to most cancers that positron emission tomography (PET) scans, which have emerged as an important tool in diagnosing cancer, works simply by revealing the places in the body where cells are consuming extra glucose. In many cases, the more glucose a tumor consumes, the worse the diagnosis.

Warburg is no longer a footnote in history, scientists

now believe that that it will be possible to slow or even stop tumors by disrupting one or more of the chemical reactions a cell uses to reproduce, and, in the process, starve cancer cells of the nutrients they need to grow.

Examining Oetzi the Iceman

Finding the body of a person who died 5300 years ago would be most rare. But it happened in 1991 when a group of hikers found a frozen corpse and his belongings in the mountainous border between Austria and Italy. The body dubbed "Oetzi the Iceman" had been entombed in an alpine glacier. (Oetzi is the name of the mountain range.) It was exceptionally well preserved due to a combination of glacial melt water and the extreme cold. Connective tissue and nervous system components were still intact, and thus Oetzi holds the record as the oldest naturally preserved ice mummy. The circumstances of his death became the world's most ancient murder case and it sparked a political and diplomatic uproar and worldwide scientific curiosity. For the former, there was an diplomatic tussle over his resting place and which country would gain access to his remains, it finally ended up in an archeological museum in Bolzano, Italy. A host of forensic pathologists and archeologists clamored for the opportunity to examine the remains. Subsequent analysis told a fascinating tale. An initial investigation suggested that Oetzi died of exposure during a winter storm and that the corpse was approximately 45 years of age, was 5'2" tall and weighed 110 lbs. CT scans unexpectedly revealed that an arrowhead lodged in his left shoulder had severed a major blood vessel between the rib cage and the left scapula. Such a wound could have led to hemmorhagic shock and his quick demise. Further examination of Oetzi's red blood cells with sensitive analytic tools confirmed that he had sustained several injuries before his death. He also had a laceration on his hand suggesting he had been fighting. This led to speculation about a fugitive fleeing an unknown lynch mob. The autopsy proved conclusively that Oetzi did not die from exposure to the cold as originally surmised but the victim of an unsolved murder that occurred in late spring or early summer around 3300 BC.

Öetzi suffered a variety of ailments, including gallbladder stones, lyme disease, whipworms in his colon, and atherosclerosis. Researchers have sequenced Öetzi's entire genome and identified a genetic predisposition to heart disease. Moreover, he had brown eyes, was lactose intolerant, and had blood type "O". There was also evidence of metal in hair samples and he might have been involved in mining or smelting copper. Additional studies continue to reveal more about the Iceman.

3D reconstructions using computer tomography offered new insights into the Iceman's oral health, showing how severely he suffered from advanced periodontitis, an oral disease that causes chronic inflammation of the tissue surrounding the teeth. The examining dentist found loss of the periodontal supporting tissue that extended nearly to the tip of the root, particularly in the rear molar area. The tooth decay can most likely be attributed to Oetzi eating more and more starchy foods such as breads and cereal porridge. These food items were more commonly consumed in the Neolithic period than earlier because of the rise of agriculture. The Iceman's abraded teeth surfaces demonstrated the abrasive nature of his food, probably due to contaminants and the rub-off from the quern (a mill used to grind grain). Oetzi also sustained mechanical damage to some teeth, which along with his other injuries testify to a tough life. One front tooth suffered damage with discoloration still clearly visible, and one molar lost a cusp, probably from chewing on something akin to a small rock in the porridge.

Oetzi was tattooed, and offers the earliest direct evidence that tattooing was practiced in Europe by at least the Chalcolithic period (5500 to 3000 BC). However, until now it has been difficult to conclusively catalog all of his marks. Oetzi's epidermis naturally darkened from prolonged exposure to sub-zero temperatures as he lay beneath the glacier, and as a result some of his tattoos became faint or invisible to the naked eye. Consequently previous studies have identified between 47 and 60 tattoos on the Iceman's body.

There was a recent report from researchers who conducted the first in-depth analysis of the iceman's stomach contents. The study offers a rare glimpse of his ancient eating habits. Among other things, their findings show that the Iceman's last meal was heavy in fat. The researchers combined classical microscopic and modern molecular approaches to determine the exact composition of the Iceman's diet prior to his death. Analysis identified ibex (wild goat) and red deer tissue as the most likely sources of fat. In fact, about half of the stomach contents were composed of animal fat. While the high fat diet was unexpected, the researchers say it totally makes sense given the extreme alpine environment in which the Iceman lived and where he was found. Fat makes an excellent energy source for anyone exposed to cold weather. Unfortunately, high fat intake has a strong correlation with increased risk of coronary heart disease.

Oxygen

—The subject of this year's Nobel Prize

Oxygen was discovered almost 250 years ago, but there is still much to be learned about this fundamental life giving element. Research is still ongoing and scientists continue to provide more information about its properties. Last week the Nobel Prize in Physiology or Medicine was awarded to three of them. These physicians shared the prize for their studies on how cells in the human body respond to low oxygen levels. According to cellular biologist Celeste Simon of the University of Pennsylvania Perelman School of Medicine, "oxygen limitation is a part of virtually all diseases, not just solid tumors or stroke, but inflammation, wound healing, and peripheral arterial disease. All of these involve decreased oxygen. The trio's work revealed the mechanism for one of life's most essential adaptive processes and promises powerful new treatments for diseases, including anemia, heart attacks, strokes and cancer.

Discovery

Before discussing the need for oxygen, it may prudent to know more about the controversies surrounding the discovery. In 1772, Carl Wilhelm Scheele found something he called "fire air", while independently, in 1774, Joseph Priestly allegedly discovered "dephlogisticated air" (phlogiston being a substance supposedly given off in burning—a reverse of oxygen). In 1777 Antoine Lavoisier published a new theory of combustion which clarified the role of the new gas, he named "oxygen", meaning (from the Greek), "acid-producer, because he mistakenly thought it to be an essential component of all acids. (It isn't course, hydrochloric acid is one example.) The nature of acids was not truly clarified until Sir Humphrey Davy's work in 1812. Even Lavoisier did not understand oxygen properly; discovery is often an extended process, and one that can be identified only retrospectively. In the case of oxygen it can be said to have begun in 1772 and ended only in 1812.

CLASSIFICATION

Oxygen is element No.8 in the Periodic Table of Elements, it is a colorless, odorless and tasteless gas that makes up 21 percent of the Earth's atmosphere. It is the most reactive of the non-metallic elements. Earth has been oxygenated for about 2.3 billion years, and levels began to creep up at least 2.5 billion years ago, according to a 2007 NASA-funded study. No one knows quite why this gas suddenly became a significant part of the atmosphere, but it is possible that geologic changes on Earth led to oxygen produced by photosynthesizing organisms.

Oxygen defies easy classification. Ever since it was discovered, its properties and chemistry have been squabbled over by scholars and charlatans alike. The controversy persists today. Oxygen is hailed as the Elixir of Life—a wonder tonic, a cure for ageing, a beauty treatment and a potent medical therapy. Although the fundamental importance of oxygen has been understood for centuries, little, however, is known about cellular response to changes in oxygen levels. This is critical to understanding how oxygen levels affect metabolism and physiological function. The recent work done to obtain the Nobel Prize will help.

Significance

As mentioned earlier everyone agrees that without oxygen there is no life, at least on this and probably every other planet. If we stop breathing it, we'll be dead in minutes. Our bodies are beautifully designed to deliver oxygen to each of our 15 million million cells. According to Nick Lane in his book, *Oxygen*: "All the symbolism of red blood ultimately rests in the simple chemical bonding between oxygen and hemoglobin in our red blood cells. Suffocation and drowning—the physical deprivation of oxygen—are among the darkest of human fears. If we think of a planet without oxygen, we think of a sterile place pockmarked with craters, a place like the Moon or Mars. The presence of oxygen in a planetary atmosphere is the litmus test of life: water signals the potential for life, but oxygen is the sign of its fulfilment—only life can produce free oxygen in the air in any abundance. If pressed for an unemotional reason for not cutting down the rain forests or polluting the oceans, we may argue that these great resources are the 'lungs' of the world, ventilating the Earth with life-giving oxygen." This is not true, as we shall see, but illustrates the reverence we hold for oxygen. Perhaps it is not surprising that we seek mystical or healing properties from this gas.

Final thoughts

As Bill Bryson comments in his latest book, "Oxygen is the biggest component in the human body, filling 61 percent of available space. It accounts for about two-thirds of our composition. The reason we are not light and bouncy like a balloon is that the oxygen is mostly bound up with hydrogen (which accounts for another 10 percent of you) to make water—and water, as you will know if you have ever tried to move a wading pool or just walked around in really wet clothes, is surprisingly heavy."

Oxytocin

—*A surprisingly versatile hormone.*

HORMONES ARE CHEMICAL messengers that are secreted directly into the blood, which carries them to organs and tissues of the body to exert their functions. They control growth, metabolism, behavior, sleep, lactation, stress, mood swings, sleep–wake cycles, the immune system, mating, fighting, fleeing, puberty, parenting, and sex. They aim to get us back to normal when things are out of whack and can be the cause of commotion, too. There are many types of hormones that act on different aspects of bodily functions and processes. Scientists continue to learn more about them and current research has uncovered new information about one hormone in particular. Prior to these studies oxytocin was primarily used to initiate and improve uterine contractions in child birth. New studies have shown that oxytocin can affect maternal behavior, social bonding, and even sexual pleasure. Moreover, oxytocin can make us more sympathetic, supportive and open with our feelings such that it extends into maintaining romantic relationships. There have also been research done to find out whether oxytocin can be used to correct some of the interpersonal deficiencies brought on by autism. Books have been written extolling the properties of oxytocin, calling it the hormone of calm, love, trust, passion and intimacy. It along with dopamine, endorphin, and serotonin has been described as one of the happy chemicals. Oxytocin is even being tested as an anti-anxiety drug. One website gives ten reasons why oxytocin is the most amazing molecule in the world, perhaps not an exaggeration.

Oxytocin

The word oxytocin was derived from Greek *oxys*, and *tokos*, meaning quick birth, after its uterine-contracting properties were discovered by British pharmacologist Sir Henry Hallett Dale in 1906. The nine amino acid sequence in its structure was discovered by Vincent du Vigneaud and by Tuppy in 1953 and synthesized biochemically soon thereafter by duVigneaud that same year. Oxytocin is a very abundant neuropeptide exerting a wide spectrum of central and peripheral effects as a neurohormone, neurotransmitter, or neuromodulator. It is produced mainly in the hypothalamus, where it is either released into the blood via the posterior lobe of the pituitary gland, or to other parts of the brain and spinal cord, where it binds to oxytocin receptors to influence behavior and physiology. The behavioral effects are thought to reflect release from centrally projecting oxytocin neurons, different from those that project to the pituitary gland. Oxytocin receptors are expressed by neurons in many parts of the brain and spinal cord. Peripheral actions mainly reflect secretion from the pituitary gland and are responsible for the stimulatory effects on the mammary glands (the hormone activates contractions of muscular cells around the milk glands) and the uterus.

POTENTIAL USES

The excitement over oxytocin began in the 1990s when researchers discovered that breast feeding women are calmer in the face of exercise and psychosocial stress than bottle feeding mothers. More recent literature has shown a number of other uses. Oxytocin levels are high under stressful conditions, such as social isolation and unhappy relationships.

New investigations show that oxytocin can make us more sympathetic, supportive and open with our feelings. These findings have led some researchers to investigate whether oxytocin can be used in couple therapy. Oxytocin levels increase during couple's periods of falling in love and it also correlated with the longevity of a relationship. A recent study was the first to assess whether people with variations in the their oxytocin receptor gene have a harder time maintaining romantic relationships than those who do not. The researchers found that women with a specific variation weren't as close to their partners as women without it. These women were more likely to report having had a marital crisis. Although it is not known how this variation affects the oxytocin system, it may result from fewer oxytocin receptors in the brain. People with fewer receptors would be less sensitive to the hormone's effects.

Oxytocin appears to play a dual role in triggering or reducing anxiety, depending on the social context. A recent study has linked oxytocin to social stress and its ability to increase anxiety and fear in response to future stress. Oxytocin appears to strengthen negative social memory and future anxiety by triggering an important signaling molecule that becomes activated for several hours after a negative social experience. Experiments with mice have established that oxytocin is essential for strengthening the memory of social interactions and that it increases fear and anxiety in future stressful situations.

Individuals with autism show altered face processing and brain activations to facial stimuli. In healthy adults, oxytocin promotes both and modulates brain activity. Magnetic resonance imaging has shown that a single intranasal dose of oxytocin increases amygdala activity in response to facial stimuli in autistic adults. This suggests that oxytocin might be used as a treatment modality. (Individuals with autism fail to recognize faces and to integrate facial expressions with emotions caused by impaired social cognitive abilities.)

The Pancreas

—Mysteries of a Hidden Organ

Many years ago, one of America's most accomplished medical writers, Dr. Lewis Thomas, described how insecure he would be if faced with taking on the responsibilities of his liver. He wrote that he was considerably less intelligent and unable to make all of the decisions the liver is designed for. Dr. Thomas would have been even more humble had he chosen the pancreas. When it functions properly, this organ is silent and unrecognized, but when it malfunctions, extraordinary complications and even death can ensue. Even now, in the 21st century when medical discoveries abound, the pancreas carries an aura of mystery and its diseases present a challenge for diagnosis and treatment. Fortunately, research is ongoing and progress is being made to treat pancreatic diseases.

The pancreas was first described by the Greek anatomist and surgeon Herophilus, in approximately 310 BC. Herophilus (335 B.C. to 280 B.C.) may have been one of the first surgeons to dissect the human body using criminals from prison. He was given royal permission to procure them, even though there was public criticism of the practice. The reason may have been that some of his dissections were performed while the criminals were still alive. Herophilus, however, did not name the pancreas. The term may first have been used by Aristotle (384–322 BC) in his treatise *Histora Animalium*, although there was some controversy regarding his description.

It was not until the latter half of the fifteenth century that the first signs of advancements in medicine became apparent. Giacomo Berengario da Carpi (1470–1550), a professor of surgery in Bologna, Italy, produced the first illustrated textbook of anatomy. He accurately described the bile duct and wrote of the pancreas as a secretory gland. His work was followed by one of true genius several years later. In 1543, Andreas Vesalius (1515–1564) published his famous textbook of anatomy, *De Humani corporis fabrica libri VII* ("Seven books on the workings of the human body"). With its publication, medicine was finally lifted out of medieval murkiness and misinformation. The book, an outgrowth of the vigorous spirit of the Renaissance, corrected a number of Galen's errors and returned society to the logical thought and observational methods of the ancient Greeks.

The pancreas is located partially behind the stomach and attached to the duodenum (a portion of the small intestine connected to the stomach). The human pancreas is a large gland that weighs about 70 grams. It is about six inches long, and its inverted, bowl-shaped head and elongated body make it look like a salmon-colored snake lying transversely across the back of the upper abdomen. The pancreas is composed of two main elements: exocrine tissue and endocrine tissue. Exocrine tissue makes up the bulk of the pancreas. It is organized into a large number of saclike structures whose interiors are lined with cells that secrete various enzymes important to the digestive process. As these pancreatic juices are made, they flow into the main pancreatic duct.

Scattered throughout the exocrine tissue are small, isolated pockets of endocrine tissue; these pockets are known as the islets of Langerhans. The islets are composed of several types of cells that produce insulin and glucagon. Insulin helps control carbohydrate metabolism, while glucagon counters the action of insulin.

Researchers today remain intrigued by the mysterious means by which the pancreas shields itself from harm. It can churn out huge quantities of enzymes to rapidly reduce our fast food diets into particles of amino acids, carbohydrates and fats without digesting its own tissue in the process. Researchers suspect that the organ's self-protective mechanisms come with a terrible downside that helps explain why pancreatic cancer can be so difficult to treat.

Pancreatic cancer is one of the deadliest of all solid malignancies. The five-year survival rate is only 4%. More than 35,000 Americans were diagnosed with pancreatic cancer in 2007, making the disease the fourth leading cause of cancer deaths. Pancreatic cancer is rare before age 40; the median age at diagnosis is 73. Cigarette smoking is by far the leading preventable cause of pancreatic cancer; it doubles the risk. It is believed that as many as one in four cases may be attributable to smoking. Other established risk factors include a diet high in meats and fat, low serum folate levels, obesity, longstanding diabetes mellitus, chronic pancreatitis and a family history of the disease.

The pancreas is also the linchpin in two current epidemics. When the pancreas goes awry it is the source of diabetes, which afflicts more than 23 million people in the US. And as the tireless brewer of digestive juices, the pancreas is at the front line of the growing incidence of obesity that affects 60% of Americans. Researchers have found that the pancreas helps mediate appetite-related messages between the brain and the gastrointestinal tract. However, it is not quite clear how the pancreas conveys sensations of hunger and satiety. When these are understood, the knowledge could lead to new ways to combat obesity.

Passenger Pigeons

—*What happened?*

When someone reaches my age a would be comedian might ask what it was like when dinosaurs roamed the earth. Of course, most anyone knows that these huge reptiles vanished more than 65 million years ago likely due to an asteroid or meteor striking the earth with deadly climatic and geological changes. However, should that same person inquire about passenger pigeons, that is another matter. While I was not witness to their demise, my mother was, and so was anyone else alive in 1914. Passenger pigeons were once the most abundant birds in the world, with an estimated population of at least 3 billion in the 1800s. By 1902, none survived in the wild, and on September 1, 1914, the very last one, named Martha, was found dead on the floor of her cage in the Cincinnati zoo. There was a concerted effort after that time to search for survivors and rewards were offered to anyone finding a nest or nesting colony, but these attempts were futile. The same was true with efforts to breed the early surviving birds. (Passenger pigeons require large numbers for optimum breeding conditions.) The extinction in a mere five decades is a reminder that even a bird numbering in the billions can be decimated within a human lifetime.

In the 1800s and before, migratory flocks of passenger pigeons were so immense that they blanketed the skies of eastern North America. One individual recounted a mile wide swath passing overhead blocking the sun for three consecutive days. The birds flew at an estimated speed of about sixty miles an hour. Nesting birds took over entire forests, trees were crammed with dozens of nests, collectively weighing so much that branches would break and tree trunks would topple. Surface vegetation was destroyed by the thick layers of the bird's excrement. Sound was overwhelming. But the birds were not just noisy, they were tasty too, and their arrival guaranteed an abundance of free food. Unfortunately, passenger pigeons also devoured crops, caused great loss to farmers, and greatly influenced forest composition by consuming and dispersing acorns, beechnuts, seeds and berries. Worms and insects supplemented their diet in spring and summer.

The scientific name for passenger pigeons is *Ectopistes migratorius*. *Ectopistes* means "moving about or wandering" and *migratorius* means migrating. The name indicates that the bird migrates not only in the spring and fall, but one that also moves about from season to season to select the most favorable environment for nesting and feeding. The bird was designed for grace, speed and maneuverability, the wings were powered by large breast muscles that provided the capability for prolonged flight.

Of course, other species have become extinct, but the passenger pigeon is the most famous example, and thus the species has been extensively studied. Human exploitation can account for most of the loss of these magnificent birds by reducing their breeding grounds and relentless hunting. James Audubon, the famous naturalist, noted on his travels along the Ohio river, that as soon as the flocks of birds arrived, the river banks were crowded with men and boys shooting at them incessantly, and multitudes were destroyed. Throughout the 1800s the hunt for passenger pigeons grew until it became a massive slaughter. New DNA studies, however, indicate that humans were only the final straw in destroying a species that was already vulnerable and headed for trouble. Research involving the pigeon's genome has shown that the total number of passenger pigeons had widely fluctuated throughout the last million years. The population was already in decline and plummeting long before hunting increased. Unfortunately, the human element occurred during the decline phase. One other factor was the reduced availability of seed crops for feeding that occurred as forests were converted into farms.

As mentioned above, other animals including, the dodo, the great auk, the thylacine (Tasmanian tiger), the Chinese river dolphin, and the imperial woodpecker have been driven extinct. Many more species are now endangered. An example is the saiga, a type of antelope, that recently suffered from a mysterious illness which killed more than half of the species in less than a month. Because of impending extinction, there is now an effort to bring some of these species back to life. One animal brought back, only to become extinct again, was a wild goat known as bucardo. It was a large creature, weighing up to 220 pounds with long, gently curving horns. Hunters, however, drove down their population over several centuries. By 1989, there were just a dozen or so left. The last one, a female, had its cells preserved, and used to implant eggs into surrogate goats. Only one survived, a 4 pound hybrid that died shortly after birth.

It is highly unlikely, that such an effort will be made to de-extinct the passenger pigeon. How would residents of New York or Chicago feel about droves of new pigeons arriving in those cities, darkening the skies and depositing their waste over the streets and buildings?

Patient counseling

—Ask your pharmacist!

When I was a practicing pharmacist in the 1950s through the early 1990's it was not uncommon to discuss the proper use of medications with a patient. As a matter of fact, pharmacists actually waited on the patient and even if not requested, discussed the proper means of administration, possible side effects and drug interactions. There was no internet around as a source for information. Today, with the proliferation of busy chain and big box stores, pharmacists are not as accessible and are usually found isolated standing behind computers in the back of the store. The current setup acts as a barrier to receive information. While it is fair to say that a patient can ask the pharmacist for advice, it is less common than in the old days and fewer independent stores, if any, are around. Even with this set up, patients should seek the knowledge and skills offered by pharmacists. As such they can provide effective and accurate patient education and counseling. One reason is that pharmacists are better educated than graduates of yesteryear. Courses are offered today that were not available years ago, pharmacokinetics is one example. **Best of all, unlike other professions, the advice from a pharmacist comes without charge. Most important, is that**

pharmacists can contribute to positive outcomes by motivating patients to follow their drug regimen and monitoring plans.

Theoretically, pharmacists should know about their patient's cultures, especially health and illness beliefs, attitudes, and practices. A trained pharmacist will be aware of patients view on their own roles and responsibilities for decision making and managing their own care. The pharmacist will assess a patient's cognitive abilities, learning style, and sensory and physical status to provide the information that meets the patient's needs. A pharmacist needs to determine whether a patient is willing to use a medication and whether he or she intends to do so. Today, many pharmacists have a Doctor of Pharmacy degree (Pharm. D.), meaning that they have attended an accredited university for at least six years and serve time working as an apprentice in a drug store or hospital. There are also continuing education requirements. It is a good practice to choose your pharmacist as you would your doctor. For example, a well read pharmacist would be aware of new reports suggesting that patients take antihypertensive medicine at bedtime rather than in the morning to improve blood pressure control. New clinical information is now readily available to disclose the best times to take other drugs as well.

Perhaps I am old fashioned but I am saddened by the loss of independents. In 2011 there were 23, 106 independent pharmacies, by 2017 that number dipped to 21, 909 despite an increase in actual population growth. The number of independents is expected to fall below 20,000 before 2025. Many large retailers are making attractive offers to independent owners to sell their businesses and many independent operators are older and looking for an exit, while younger pharmacists are less attracted to the risk and return of ownership. Pill Box pharmacy here in Warsaw has addressed some of that loss by setting up a satellite or telepharmacy in Albion, a town without a drug store. It was the first of its kind in Indiana and required a new state law to allow that to happen. The Albion location is deemed a remote dispensing facility, set up like a normal pharmacy. Albion was without a pharmacy since December, 2006, when CVS transferred all of the prescription files to a store in Kendallville and closed. Pill Box will use image capture and real time video conferencing to bring retail pharmacy back to Albion. Using telepharmacy software Pill Box will share a full time pharmacist between two locations, giving the residents of Albion access to his or her counseling and the services of a pharmacy technician. The technician makes an original prescription available to the pharmacist here in Warsaw by placing it in view of the video communication system or by scanning the prescription into the shared electronic record keeping system. The new technology stores all of the transaction and helps to minimize the possibility of errors. Patient counseling is mandatory when new drugs are dispensed.

Final thoughts

A pharmacist is the key to proper drug usage. This includes the expected benefits and action and whether the medication is intended to cure a disease, eliminate or reduce symptoms, arrest or slow a disease process, or prevent the disease or symptom. The pharmacist can describe the action taken in case of a missed dose and potential common and severe adverse effects that may occur along with actions to prevent or minimize their occurrence. Because many people take more than one drug, a pharmacist can provide information about potential drug-drug , drug-food and drug-disease interactions. Pharmacists can also counsel patients in the proper selection of nonprescription or over-the-counter medications. Why not take advantage of this marvelous gift?

Philosophy

~~~~

—A subject to know more about

Unfortunately I did not have the opportunity in college to study philosophy or learn about philosophers and therefore had little interest in the subject. This recently changed after I discovered a small book entitled *Montaigne* written by Stephan Zweig—it captured the essence of Michel de Montaigne's life and accomplishments. Montaigne is one of history's great thinkers, and despite the fact that his essays were written in the 16th century, they continue to be quoted and widely read. Each of them provide guiding principles to enrich our lives. The format for his essays became a new form of literary expression, a brief and incomplete treatment of a topic germane to human life. His popularity has increased in recent years and Montaigne is now regarded by many as one of the most important figures in the history of philosophy. I was surprised that I knew so little about him.

An example is something he wrote about retirement which he viewed as an opportunity to pursue a more solitary life in which a person's worth is not measured in paychecks, awards, or promotions. "We have lived quite enough for others: let us live this tail-end of life for ourselves." Montaigne writes in his late 16th century essay "On Solitude." "Withdraw into yourself, but first prepare yourself to welcome yourself there." Sound advice. Other of his quotes include, "Ambition if not a vice of little people", "My life has been full of misfortunes most of which never happened" and "He who establishes his argument by noise and command shows that his reason is weak." Montaigne was once on the verge of dying after an accident and found himself gasping for air, and attempting to pound on his chest to breathe. Fortunately he recovered. He later reflected that despite the trauma, he began to grow languid while feeling like he was being carried aloft on a magic carpet. From this he found that learning to die is not necessary. He noted, "If you don't know how to die, don't worry; nature will tell you what to do on the spot, fully and adequately. She will do the job perfectly for you; don't bother your head about it."

Montaigne was meant to be a philosopher. He was born in France on February 28, 1533 and his father assured that he was raised in the most caring and careful manner. As a child he was awakened by the sound of light and agreeable music. Around his bed, flute and string players waited for the signal to play a soft melody to draw the sleeping child from his dreams. He learned Latin, without suffering the rod or shedding a tear, before beginning French, thanks to the German teacher whom his father had placed near him, and who never addressed him except in the language of Virgil and Cicero. The study of Greek took precedence. At six years of age young Montaigne went to the College of Guienne at Bordeaux, where he had as preceptors the most eminent scholars of the sixteenth century. At thirteen he passed through all the classes, and as he was destined for the law, left school to study that science. He was then about fourteen, but these early years of his life are involved in obscurity. The next information that we have is that in 1554 he received the appointment of councillor in the Parliament of Bordeaux; in 1559 he was at Bar-le-Duc with the court of Francis II, and in the year following he was present at Rouen to witness the declaration of the majority of Charles IX. He did not begin writing until he was almost 50 years old.

He wrote free floating pieces with simple titles including Of Friendship, Of Cannibals, Of the Custom of Wearing Clothes, Of Names, Of Smells, Of Cruelty, Of Diversions, Of Experience and others. Altogether there were one hundred seven such essays. Some occupy a page or two, others run much longer up to a thousand pages. All of his literary and philosophical work is contained in his essays, which he began to write in 1572, and first published in 1580 in the form of two books. Over the next twelve years leading up to his death in 1592, he made additions to the first two books and completed a third, bringing the work to a length of about one thousand pages. As time has gone by his admirers have increased exponentially and proof of his genius. Anyone with an interest in philosophy would be well advised to start with Montaigne.

They teach us the art of self-help (Cicero writes that philosophy teaches us to "be doctors to ourselves"), but self-help of the very best kind, that doesn't focus narrowly on the individual, but instead broadens our minds and connects us to society, science, culture, and the cosmos. The course is not prescriptive—the faculty don't agree with each other (in fact, some of them actively dislike each other), and the book doesn't put forward one philosophy, but several. And yet, as in Raphael's painting, there is a unity underlying the diversity: all the teachers share an optimism in human rationality, and in the ability of philosophy to improve our lives.

Poems

—Why do I find them so inscrutable?

Robert Frost, one of America's great poets wrote: "Poetry is when an emotion has found its thought and the thought has found words." Poetry is an art form that dates back to ancient times and a form of expression that makes our society more civil. That being the case, why do I have a difficult time understanding most poems I read? Why do I find it so hard to memorize them? Are these reasons why reading poetry is becoming an endangered art? My queries stem from an editorial I recent read in a medical journal. In that issue, the author quoted some lines from Cecil Day Lewis' poem, *Walking Away,* and then interpreted the contents:

> *I can see*
> *You walking away from me towards the school*
> *With the pathos of a half-fledged thing set free*
> *Into wilderness..*
>
> *That hesitant figure, eddying away*
> *Like a winged seed loosened from its parent stem,*
> *Has something I never quite grasp to convey*
> *About nature's give-and-take..*

I have had worse partings, but none that so
Gnaws at my mind still. Perhaps it is roughly
Saying what God alone could perfectly show—
How selfhood begins with a walking away,
And love is proved in the letting go.

The author of the journal article, Belinda Jack, as a means of interpretation, tells us that the poem recalls our first day at school, ourselves being the "hesitant figure". Those of us who have had children will remember their child's first day at school and the accompanying parental anxiety and the sense of loss. At the close of the poem, the lines "How selfhood begins with walking away,/And love is proved in the letting go" suggest much that is central to child development psychology. "Selfhood begins with walking away"; is according to the author, an extraordinary concise way of expressing the fact that a child has to establish independence by distancing itself from its mother and then its family. "Love is proved in the letting go" describes the other side of the equation. The parent has to accept, even encourage this separation, even if it goes against certain protective instincts and this can be painful for the adult and the child. In further discussion, the likening of the child to a "fledgling" and a "winged seed" suggests that this parting is part and parcel not just of the poet's experience, but fundamental to nature—both creatures and the botanical. The wilderness is both a literal one, nature in the wild, and a metaphor for the playground with all its unknowns. After reading the poem, again according to Belinda Jack, a great poem may give us a sense of understanding of emotions that might otherwise remain an amorphous mass of unprocessed feeling. I accept all of the explanation after reading it several times but doubt I could do the same with each poem I read. Moreover, Walking Away, unlike other poems, is not that complicated.

The solution to my dilemma lies in continuing to read books about poems, how to read them and how to understand them. There are a number to choose from and indicative of the reason such books are urgently needed. One of the best is "How to Read a Poem", written by Edward Hirsch, describes why poetry matters and helps open our imaginations as to the message the poem sends us. One much older is "Understanding Poetry" by Cleanth Brooks and Robert Penn Warren. This book is now in its fourth edition. "Essential Pleasures" edited by Robert Pinsky is an great collection of poems to be read aloud. Another excellent collection of poems can be found in Garrison Keillor's book, "Good Poems".

My favorite poem is Nobility written by Alice Cary, born on April 26, 1820 and died February 12, 1871, the first stanza is:

> *True Worth is in being, not seeming—*
> *In doing, each day that goes by,*
> *Some little good—not in dreaming*
> *Of great things to do by and by.*
> *For whatever men say in their blindness,*
> *And spite of the fancies of youth,*
> *There's nothing so kingly as kindness,*
> *And nothing so royal as truth.*

One editor ranked Nobility as one of the best and most beautiful poems every written. I continue to read it along with other great poems. With regard to my inability to memorize poems, I have included one that I have:

There once was a fellow from Yuma
Who told an elephant joke to a puma.
Now his skeleton lies under hot western skies,
The puma had no sense of huma.

Another polymath

—John Von Neumann

Polymaths are defined as humans of exceptional versatility, who excel in multiple, seemingly unrelated fields. I have also been interested in such people and continue to read about them. In the past I have written about Leonhard Euler, Thomas Young, Hermann Helmholz, Joseph Leidy and Isaac Newton. Each were extraordinary men who met the prerequisites of a genuine polymath. Newton, for example, continues to be the pinnacle used to measure everyone else. The others are equally so. The criteria I used for selection were based on contributions to science, productivity and versatility. Euler may have been the premiere early mathematician, he provided new insights into geometry, trigonometry and calculus. He was also a revolutionary thinker in astronomy, acoustics, hydrodynamics, mechanics, music, ballistics, navigation and topology and wrote more than 850 publications, including 18 books. Young was a linguist, physician and physicist who established the theory of light, color perception, anatomy, the significance of energy, elasticity and because of his uncanny knowledge of language, the study of Egyptology and hieroglyphics. Helmholtz's scientific achievements and his philosophical reflections on science ultimately became a collection of seven thick volumes, covering physiological optics, physiological acoustics and music, popular science and philosophy and lectures of theoretical physics. Leidy, described as the man who knew everything, was America's foremost microscopist and human anatomist, he was a prolific writer and lecturer, the founder of the American vertebrate paleontology, and the first to describe dinosaurs, several varieties of amoebas, plants, worms, reptilian and mammalian fossils. Of course, I could have included Leonardo da Vinci, often called the universal genius. Although perhaps the greatest artist and sculptor, he did not contribute as much to science as the individuals on my list. Leonardo had almost no schooling and could barely read Latin or do long division. Moreover, many of his manuscripts did not survive.

Proving that I have much to learn, I have been watching a series of lectures from the Great Courses on Philosophy, which includes many of the great thinkers beginning with the ancient Greeks. According to the instructor of that course, the most brilliant man lived not too long ago. He chose John Von Neumann, who died in 1957. Von Neuman was a Hungarian-American mathematician, physicist, and computer scientist. Among his awards were the Medal of Freedom, Bocher Memorial Prize, Medal for Merit and the Enrico Fermi Award. The Medal of Freedom recognized his work on developing the atomic bomb. Had Von Neumann lived longer, he undoubtedly would have won three or more Nobel Prizes. A brief search of Wikipedia and other sources indicated many of his astounding achievements.

As might be expected, Von Neumann was an extraordinary child prodigy in the areas of language and mathematics, as a six year old he could divide two 8-digit numbers in his head. By the age of eight, he was familiar with differential and integral calculus.

Apart from being a great mathematician and physicist, he made seminal contributions to computing and economics. He was a master in each field. By 1927, he was recognized as a genius, with 12 major papers in mathematics and renowned for his powers of memorization. His work in mathematics included functional analysis, geometry, topology and numerical analysis. According to what I have read, all mathematicians recognized his brilliance. One of his contemporaries, Paul Halmos (himself a gifted mathematician and luminary) once suggested, perhaps only half joking, that Von Neumann's abilities were so far above everyone else's, that Von Neumann might be a different species.

His 1932 work Mathematical Foundations of Quantum Mechanics played an important role in the development of quantum theory, and also established a rigorous framework for it. Von Neumann was actively involved in the Manhattan Project , the research and development undertaken during World War II that produced the first nuclear weapons. His main contribution was in the concept and design of the explosive lenses needed to compress the plutonium core of the atomic bomb. He was in the selection committee that chose Hiroshima and Nagasaki as the targets. His contributions to the field of computing are legendary. He wrote the sorting program together with Alan Turing, a precursor for Artificial

Intelligence and contributed to the Monte Carlo method which allowed solutions to be approximated using random numbers. Von Neumann was also involved in developing linear programming, self replicating machines and stochastic computing. It is mind boggling to believe that one person could contribute as much as he did to science and mathematics.

Final thoughts

There may not be a true analytical measurement for a genius. Historian Roy Porter, noted that for such individuals, "there seems to be no common denominator except uncommonness ." That is certainly true for John Von Neumann and others I have written about.

POLYPILL

—A means to live longer?

"Drugs don't work in patients who don't take them."
C. Everett Koop, M.D.

THERE IS LITTLE good news these days about Iran and its nuclear ambitions and terrorist activities with at least one exception. A recent clinical trial conducted in that country indicated that an inexpensive combination drug could lower the rate of heart attacks by more than half and help all of us live longer. The pill, together with other suggestions like eating a balanced diet, consuming less fat from meat and dairy, focusing on fruits and vegetables, practicing daily exercise, nonsmoking, reducing stress, and having routine medical checkups could be significant. The major reason is that following this regimen can prevent the major causes of premature death—-heart disease and stroke. Some experts have long believed that such life style changes should include taking multiple drugs. It may even be more advantageous to combine all of the recommended drugs into a single pill—a so called " magic bullet" to extend life expectancy. Remarkably, there has been and is a product that may meet this requirement. Unfortunately it is not available in the United States. It is, however, well known in England where it is described as the "polypill." The combination drug was first proposed in 2003 by cardiologists Nicholas Wald and Malcolm Law of Queen Mary, University of London. They studied and summed up the preventive effects of several drugs and concluded that combining the low doses of all six into a once a day pill would lower cholesterol and blood pressure. According to Wald and Law, the combination of drugs would decrease the incidence of cardiovascular disease in at-risk patients by up to 80 percent.

The Iranian study was much larger, it involved the participation of 6,800 rural villagers aged 50 to 75 and contained a cholesterol lowering statin, two blood pressure drugs and a low dose aspirin . It was conducted by doctors from Teheran University, the University of Birmingham in Britain and other institutions. Fortunately it was the first study of such a multidrug pill that was large and long-lasting enough to measure "clinical outcomes" including how many people actually had heart attacks, strokes, or episodes of heart failure while taking the pills, rather than just how many, for example, lowered their blood pressure or cholesterol. This study has greater significance as more residents of poor countries survive childhood into middle age and beyond—and as rising incomes contribute to their adoption of cigarette smoking and diets high in sugar and fat. About 18 million people a year die of cardiovascular disease, and 80 percent of them are in poor or middle income countries threatened by rising rates of obesity, diabetes, tobacco use and sedentary living. Similar studies are underway in many countries.

Constituents

The polypill used in the Iranian study included aspirin, atorvastatin (to reduce cholesterol), hydrochlorthiazide (a diuretic), and either enalapril (an ACE inhibitor) or valsartan (an angiotensin receptor blocker). Aspirin is used for its effects on platelets to reduce blood clots. Thiazides are diuretics that stimulate the flow of urine. Angiotensin converting enzyme (ACE) inhibitors and angiotensin receptor blockers lower blood pressure and help keep blood vessels open.

The pitfalls

Despite the good news from Iran there may be certain drawbacks to consider. For one, patients may rely on the pill rather than adhering to a healthy lifestyle. And, of course, it is imperative that the value of the polypill be clearly demonstrated through long term clinical studies in the United States rather than simply assume that it works. The U.S. Food and Drug Administration requires evidence of efficacy in populations with low risk as well, perhaps, as evidence that each component of the polypill adds something important. There are side effects to consider as well, i.e., aspirin induced bleeding. In the U.K. study mentioned above about 1 in 6 patients experienced a side effect in the short term. Most were mild but about 1 in 20 overall stopped treatment due to side effects, indicating that treatment is best targeted to those at raised risk of disease. Side effects may take five to seven years to emerge. Even though the drugs used are generic, there are doubts as to whether developing countries could afford to provide them broadly to everyone over

55 years of age. Cardiologists are critical of the one-size-fits-all treatment of patients who may not be at risk. Many physicians want to be involved in personalized care. The availability of the polypill may keep patients away from doctors for routine examinations. Not all patients are ready to assume self care and autonomy.

Final thoughts

According to the Science News Daily, the polypill will be available soon in India and then elsewhere within a few years, based on regulatory timelines within each individual country. That may not be true in the United States where combination drugs are more difficult to gain approval.

Potatoes

—Are they becoming an endangered species?

Do you want fries with that? This question asked in most fast food restaurants is generally answered in the affirmative and most of us love them. In the United States, potatoes are the No. 1 vegetable crop and the fourth most consumed crop in the world, behind rice, wheat and corn. Fried potatoes, however, are not a health food. A study published in 2017 in the American Journal of Clinical Nutrition found that people who ate them twice a week saw an increased risk of death. The study examined potato intake in 4,400 people between the ages of 45 and 79. By the end of the eight year study, 236 people had died. Researchers found that those who ate fried potatoes including french fries and hash browns, were more than twice as likely to have died. Despite the health related problem derived from the method of preparation, potatoes are a staple for 1.3 billion people and are nutritious. Phytonutrients in potatoes include carotenoids (Vitamin A), flavonoids (anti-inflammatory, antioxidants), and caffeic acid (antioxidant). Vitamin C is also found in potatoes, in fact, a medium sized potato contains about 45 percent of the recommended daily allowance. All of these substances may prevent or delay some types of cell damage, according to the National Institutes of Health and may help with digestion, heart health, blood pressure and even cancer prevention.

Potatoes may help lower blood pressure for several reasons and the fiber found in potatoes could help lower cholesterol by binding with it in the blood. (Fiber is found primarily in the outer potato peel.) In addition, potatoes are a good source of potassium, more than bananas. Potassium is a mineral that helps lower blood pressure, it acts to dilate blood vessels. Vitamin B6 in potatoes is critical in maintaining neurological health as are manganese, phosphorus, niacin and pantothenic acid.

Overall, potatoes cooked the right way—without heaps of butter, bacon and sour cream—are good for you. The best way is baking or microwaving. Either method causes the lowest amount of nutrients to be lost. The next healthiest way to cook is through steaming, which causes less nutrient loss than boiling. However you cook the potato try to eat the skin, it contains much of the nutrients and the majority of the fiber. Fortunately potatoes are low in calories, a medium sized potato contains only about 110. The level of carbohydrates found in potatoes makes them easy to digest and the fiber filled skin can help keep us regulated. All of the nutritional information about potatoes is favorable, however, while being fat free, they do not contain all of the 20 essential amino acids and 30 vitamins and minerals and they contain little protein.

Potatoes are not root vegetables, they are actually part of the stem of the perennial *Solanum tuberosum*. This part of the plant is called a tuber, and it functions to provide food to the leafy part of the plant. There are thousands of potato varieties, but not all are commercially available. Idaho is the top potato-producing state, but they are grown in a number of other states as well. The Inca in Peru were the first to cultivate potatoes, growing them around 8000 B.C. to 5000 B.C. In 1536, Spanish conquistadors conquered Peru, and carried potatoes back to Europe.

According to a recent article in Science, potato farmers are facing an uncertain future. Starting in Peru where potatoes have been grown for thousands of years, climate change is becoming a problem. Drought and frost are striking more often. Rains come later, shortening the growing season. Warmer temperatures have allowed moths and weevils to encroach from lower elevations.

To find potatoes that can cope with those challenges, researchers and Peruvian farmers are testing of the 4350 locally cultivated varieties. In Peru and around the world, enhancing the potato has become a high priority. Unfortunately, creating a new potato variety is slow and difficult, even by the patient methods of plant breeders. The reason is that commercial varieties carry four copies of each chromosome, which forces breeders to create and test hundreds of thousands of seedlings to find just one with desired combination of traits. To breed a better potato, it helps to have plenty of genetic raw material at hand. But the world's gene banks are not fully stocked with the richest source of valuable genes—the 107 potato species that grow in the wild. Habitat loss threatens many of those species. In a bid to preserve the wild diversity before it vanishes, collectors have made the biggest push ever, part of $50 million program coordinated by the Crop Trust, a charity based in Germany. The key to a robust potato may be waiting in the wild species that grow from southwestern North America through central and South America. Bringing some of the ancient diversity back into cultivation could be the answer to environmental changes to the wonderful but threatened potato.

Probiotics and Microflora

Max Sherman, RPh
President, Sherman Consulting Services, Inc.
Warsaw, Indiana

PROBIOTIC FOODS HAVE recently become popular in the United States, although such products have been marketed for decades in Europe and Asia. Probiotics are defined as living organisms that, when administered in sufficient numbers, are beneficial to the host. One example is Activia. It is the name for a line of yogurt with *Lactobacillus, Streptococcus thermophilus*, and *Bifidobacterium animalis* bacteria, and it is advertised to aid regularity. While new to the U.S., it has been sold in Europe since 1987. Most probiotic products can be found in the dairy case of supermarkets or as dietary supplements. There are probiotic frozen yogurts and dairy-based drinks such as DanActive, a probiotic yogurt drink which contains *Lactobacillus casei immunitas* cultures. Its manufacturer (Dannon) indicates that the product is clinically proven to "help strengthen your body's defenses." Products sold in the pharmac, include Culturelle (*Lactobacillus GG*), Lactinex (*Lactobacillus acidophilus, Lactobacillus bulgaricus)* and Florastor (*Saccharomyces boulardii)* yeast, which are indicated to reduce the chance of developing diarrhea due to antibiotics.

The growth of probiotics comes as many scientists are now focused on the role of beneficial bacteria to aid digestion, boost natural defenses, and fight off harmful bacteria that could cause health problems. Intestinal bacteria can benefit health by breaking down toxins, synthesizing vitamins, and defending against infection. They may also play a role in such diseases as peptic ulcers, colorectal cancer, and inflammatory bowel disease. This article will describe the genesis and evolution of our indigenous microbial community, the size and makeup of its inhabitants, their effects, benefits, and new research.

Genesis and Evolution

Most of us are aware that bacteria are a part of a healthy human ecosystem. (An ecosystem is an assembly of species and the organic and inorganic constituents characterizing a particular site.) According to one author, the armies of bacteria that sneak into our bodies the moment we are born are the "primal illegal immigrants." Most are industrious and friendly, minding their own business in tight-knit, long-lived communities, doing the grunt biochemical work we all rely on to stay alive.[3] The ecosystem forms at birth, but the human-microbe alliance begins months before. Midway through pregnancy, a hormonal shift directs the cells lining the vagina to begin stockpiling sugary glycogen, the favorite food of sausage-shaped bacteria called lactobacilli. By fermenting the sugar into lactic acid, these bacteria lower the pH of the vagina to levels that discourage the growth of potentially dangerous invaders.

The infant mouth's first inoculation of bacteria includes a generous sampling of the lactobacilli present in the mother's birth canal. With the first gulp of breast milk, these lactobacilli are joined by millions of bifidobacteria, a related group of acid-producing microbes. The source of these bacteria are the mother's nipples, where the bacteria appear during the eighth month of pregnancy. Bifidobacteria secrete acids and antibiotic chemicals which repel potentially dangerous organisms including *Staphylococcus aureus*. Bifidobacteria and lactobacilli are soon joined with acid-tolerant *Streptococcus salivarius* that appear on a baby's tongue during the first day of life. Bifidobacteria are anaerobic, pleomorphic rods which break down dietary carbohydrate and synthesize and excrete water-soluble vitamins. Their name is derived from the observation that they often exist in a Y-shaped or bifid form. These organisms predominate in the colons of breast-fed babies and account for up to 95% of all culturable bacteria and protect against infection. Strangely, they do not occur in such high numbers in adults.[6] Several other streptococci along with one or more kinds of *Neisseria* bacteria settle in during the first week. The vast majority emanate from the mother's mouth, which is always within reach of a nursing baby's fingers.

As the baby begins nursing or drinking formula, the bacterial population inside the mouth increases. These bacteria consume enough oxygen, creating a zone where anaerobic bacteria can thrive. By the time the baby is 2 months old, a microscopic close-up of the gums will reveal clusters and chains of bacteria and fungi. Another wave of bacteria arrive when the first teeth appear. The first is *Streptococcus sanguis* followed by *Streptococcus mutans*. By middle childhood, the diversity inside the mouth surpasses a hundred species, and their total number is greater than 10 billion. Bacteria also settle in the nasal cavities, which are connected to the mouth via the upper respiratory tract. The bacteria eventually lodge in the intestinal tract. In the small intestine, incoming microbes engage the infant's dormant immune system. Pits on the surface of Peyer's patches capture passing bacteria where they are ushered into the underlying lymph tissue. Interaction on the Peyer's patches triggers the production of an abundance of immunoglobulin A (IgA) antibodies. Instead of marking the bacteria for destruction, IgA clusters across the bacterial surface and keeps it from attaching to the intestinal wall. This also leads to the proliferation of T and B cells that will marshal an attack against these same bacteria should they turn up in the blood or in other forbidden territory. The small intestine must provide a platform for nutrient absorption, but at the same time, the epithelium and its associated immune cells must keep out pathogens that escape the inhospitable environment of the stomach. To satisfy these responsibilities, small intestinal epithelial cells divide at a rate of 13 to 16 cells every hour.

When the child grows up to become an adult, his or her intestine is home to an almost inconceivable number of microorganisms. The size of the population—up to 100 trillion (a trillion seconds in time would be 32,000 years)—far exceeds all other microbial communities associated with body's surfaces and more than 10 times greater than the total number of our somatic and germ cells. (There is a significant variation in both the total number of bacteria and the composition of the bacterial flora in different body regions.) Since humans depend on their microbiome for various essential services, a person should really be considered a superorganism, consisting of his or her own cells and those of all the commensal bacteria.

Humans are not inherently endowed with a healthy immune or digestive system. Fortunately, our intestinal tract, which includes our inhabitants (microbiome), provides us with genetic and metabolic attributes we have not been required to evolve on our own, including the ability to harvest otherwise inaccessible nutrients and to modify host immune reactivity.

INHABITANTS

The adult human gastrointestinal (GI) tract contains all three domains of life—bacteria, archaea, and eukaryotes. Archaea are a group of prokaryotic and single celled microorganisms, and while similar to bacteria, evolved differently. Archaea were originally described in extreme environments but have since been found in all habitats including the digestive tracts of animals such as ruminants, termites, and humans. Eukaryotes are organisms whose cells contain a limiting membrane around the nuclear material (the nucleus). Bacteria living in the human gut achieve the highest cell densities recorded for any ecosystem. The vast majority belong to two divisions, the *Bacteroidetes* (48%) and the *Firmicutes* (51%). *Bacteroidetes* include a number of *Bacteroides* genera, which have yet to be encountered in any environment other than animal GI tracts. *Firmicutes* include the genera *Clostridium*, *Lactobacillus*, *Eubacterium*, *Ruminoccus*, and several others. In the first comprehensive molecular survey of the gut microbiota (normal microflora), 395 bacterial and one archaeal phylotype (bacteria defined by their ribosomal RNA gene sequence) were identified. Thus, the gut microbiota is a tremendously diverse bioreactor. Eight divisions with divergent lineages are represented. This diversity is desirable for ecosystem stability. There appears to be a strong host selection for specific bacteria whose behavior is beneficial to the host. Cooperative activity by bacteria is required to break down nutrients and provide the host with energy. Populations of bacteria are remarkably stable within the human gut, which implies that mechanisms exist to suppress undesirable bacteria and promote the abundance of those that are needed.

Bacteroides thetaiotaomicron is the prominent and remarkable bacterial species in the distal intestinal tract of adult humans. It is a very successful anaerobic glycophile ("sugar loving" microbe) whose prodigious capacity for digesting otherwise indigestible dietary polysaccharides is reflected in its genome. It encodes 241 glycoside hydrolases and polysaccharide lyases. This means that the organism has the ability to break down xylan-, pectin- and arabinose-containing polysaccharides that are common components of dietary fiber. When dietary polysaccharides are scarce, *B thetaiotaomicron* turns to host mucus by deploying a different set of polysaccharide-binding proteins and glycoside hydrolases. Other *Bacteroides* species include *B vulgatus, B distasonis*, and *B fragilis*. All play a role in the digestive process.

NEW RESEARCH

Microbiologists from Louis Pasteur (1822-1895) and Ilya Mechnikov (1845-1916) to present day scientists have emphasized the importance of understanding the contributions of our microbiota to human health and disease. Mechnikov, who won the Nobel Prize for Physiology and Medicine in 1908, was one of the first researchers to study the flora of the human intestine. He developed a theory that senility is due to poisoning of the body by the products of some of these bacteria. To prevent them from multiplying, he suggested a diet containing milk fermented by bacilli, which produce large quantities of lactic acid.

Today, science is on the verge of understanding how the body maintains a state of equilibrium with its incredibly complex enteric microflora. Appropriate immune recognition is also essential to host–bacteria symbiosis. (Symbiosis is the biological association of two individuals or populations of different species.) It has recently been shown that the recognition of commensal bacteria by epithelial cells protects against intestinal injury. Other research indicates that use of antibiotics reduces the capacity of intestinal microflora to metabolize phytochemicals into compounds that may protect against cancer. However, antibiotic use also disrupts the intestinal microflora metabolism of estrogens, which results in lower levels that might decrease the risk of some hormonal cancers. Use of antibiotics may be associated with cancer risk through effects on immune function and inflammation, although little is known about these mechanisms.

Intestinal bacteria release chemical signals recognized by specific receptors—called toll-like receptors (TLR)—of the innate immune system. The interaction helps to maintain the architectural integrity of the intestinal surface and enhance the ability of the epithelial surface to withstand injury. A deficiency in any of the numerous signaling molecules can induce intestinal inflammation, which may be a precursor of inflammatory bowel disease. Research is now ongoing to understand various types of TLR activation to ascertain how this information can be used to treat irritable bowel syndrome, Crohn's disease, and other types of intestinal inflammatory conditions.

A group of medical researchers in Ireland recently identified five probiotic bacteria than can prevent Salmonella infection in pigs and if translated to humans could potentially reduce Salmonella-induced foodborne illnesses, which cause between 500 and 1,000 deaths every year in the U.S. This same group is also investigating the human microbiome for antimicrobials against pathogens. They have isolated a compound called lacticin 3147 from the harmless bacterium *Lactococcus lactis*, which has been used to make cheese. Recently, lacticin 3147 has demonstrated antimicrobial activity against a range of genetically distinct *Clostridium difficile* strains isolated from the human gut. This indicates that lacticin 3147 may offer a new treatment for *C difficile*-associated diarrhea, a serious condition that affects approximately 3 million people per year in the U.S. and is a major problem in hospitals.

There is evidence confirming the effects of *Lactobacillus GG* in preventing diarrhea and atopy in children. These organisms are thought to occupy binding sites in the gut mucosa that prevents pathogenic bacteria from adhering. Lactobacilli also produce bacteriocins that act as local antibiotics. Diarrhea associated with antibiotics may result from the antibiotics disrupting the normal flora in the gut of a healthy person. Such disruptions cause dysfunction of the gut's ecosystem and allow pathogens to colonize the gut and gain access to the mucosa. Whether probiotic supplements stop this process by reducing the disruption or by acting as substitutes for healthy flora is unclear.

Final Thoughts

Recent evidence has shown that microbes and their genes play important roles in the development of our immune systems, in the production of fatty acids which enhance healthy intestinal cell growth, in elaborating molecules which inhibit the growth and virulence of enteric bacterial pathogens, and in the detoxification of ingested substances that could otherwise lead to cancerous cell growth or alter our ability to metabolize medicines. Pharmacists will thus become more involved in counseling patients interested in taking probiotics. It is interesting to know that in Europe probiotics are regarded as medicines and prescribed along with antibiotics. In the U.S., pharmacists can advise patients to take such products as Culturelle, Florajen, or Flora-Q while on antibiotics and for 3 to 7 days thereafter. The same products can be taken to help prevent traveler's diarrhea. They should be taken a few days before the trip and continued through its duration. Patients should be told to separate any probiotic and antibiotic doses by 2 hours to prevent the antibiotic from destroying the probiotic organisms. Immunocompromised patients should be advised not to use probiotics because of the potential for systemic infections.

Pythons

—A model to study digestion and human heart disease.

Recent news from Florida about disappearing opossums, raccoons and deer resulting from hungry Burmese pythons is alarming. The same is true for foxes and rabbits. These pythons have been found to be solely responsible for the rapid decline, and when each of these species vanish, mosquitoes must feed on rats which have miraculously survived. Rats can carry encephalitis and that puts humans at risk of acquiring the disease. This first began sometime in the 1980s when pet pythons were regarded as exotic pets. When young, a new born python is just 10 inches long, and much to the surprise of owners, those babies grow 20 times that size. As pythons grew, frightened owners began to release them into the Everglades and the numbers have continued to grow ever since. Hurricane Andrew, which destroyed a reptile breeding facility in 1992, and releasing specimens, was also partly responsible. Now Everglades National Park is teeming with giant Burmese pythons and there is no end in sight.

Burmese pythons are one of the six largest kinds of snakes in the world. They can weigh up to 200 pounds and can grow to sizeable lengths. The largest pythons are always female, they can grow from 13 to approximately 18 feet while the typically small males grow from 8 to 17 feet.

Burmese pythons are dark colored with many brown blotches bordered in black down the back. The attractiveness of their skin pattern contributes to their popularity with reptile keepers and the leather industry. They are native to a large variation of tropical and subtropical areas of the Southern and Southeast Asia. Pythons are constrictors, therefore they don't have fangs and are not venomous. Their back curving teeth are used to seize and hold their prey. Pythons have two lungs, one more than most snakes. This species lacks eyelids but they do have a thin epidermal membrane covering the eyes to protect them. Pythons have small heat pits, or holes in their upper lip which allows them to detect heat radiations from any animal nearby. They are able to smell with the aid of the "Jacobson's organ" in the roof the mouth. They dart their tongues in out of their mouths to obtain gases in the air. The gas detection method allows the python to catch its prey in light or dark conditions. Pythons do not have to eat very often and for this reason they have proven invaluable in research to study digestion and surprisingly, heart disease.

Pythons are known for their enormous appetites. Like all snakes, they are meat eaters. In a single meal, they can devour animals as big as they are. There was a recent report about a 16 foot long Burmese python swallowing a 76 pound deer. The digestive process to such large prey has made pythons a model species to study how food is digested. Within a day the internal organs can double in size and the metabolic rate, insulin production and lipid levels rise extraordinarily. The entire digestive system undergoes a massive remodeling, with rapid swelling of the intestines, production of stomach acid, and a 40 percent increase in mass of the ventricles of the heart in order to fuel the digestive process. The python's organs return to normal size in a few days and metabolism slows. The snake can then fast for months, for even a year, without losing muscle mass or showing ill effects.

Of particular interest was the method by which a gorging python expands its heart by enlarging existing cells and not by creating new ones. Despite the massive amount of fatty acids in the bloodstream following digestion there is no evidence of fat deposition in the python's heart and there is increase in the activity of a key enzyme known to protect the heart from damage. These facts have made pythons an ideal model to study heart disease and its prevention.

For anyone who contemplates purchasing a Burmese python as a pet, it is wise to ensure that the animal has clear firm skin, a rounded body shape, a clear vent (the posterior opening for urinary and fecal excretion), clear eyes, and one that actively flicks its tongue around when handled. On the other hand, a prospective buyer should remember that the snake may grow to more than 15 feet long, weigh up to 200 pounds, have bowel and bladder habits like a horse, live more than 25 years, and for whom you have to kill mice, rats, and eventually rabbits to feed. Additionally, pythons are wild animals, unpredictable and dangerous. Something for pet owners to carefully consider.

Rats

—*Built to last!*

An editorial published in a prominent medical journal more than 100 years ago called for exterminating all species of rats to stop the spread of diseases. It was noted that rats breed three or four times a year, with females breeding when about four or five months old. (The gestation period for a pregnant female rat is twenty one days.) Male and female rats may have sex twenty times a day, and a dominant male rat may mate with up to twenty female rats in just six hours. The average litter size for rats is ten or more, and thus a single pair, breeding three litters a year, would in three years have a progeny numbering up to twenty million.

For a number of reasons including their sheer numbers, rats were, and continue to be, the most despised and feared creatures that plague mankind. They carry bacteria, viruses, protozoa, fungi, mites, fleas and ticks. According to one reference, rats have been responsible for the deaths of more than ten million people in just the past century. Moreover, rats often bite youngsters and infants on the face because of the smell of food residues on the children. Approximately 50,000 people are bitten by rats every year. Despite their frightening and abhorrent characteristics, rats have a redeeming quality not described in the editorial written a century ago. For example, rats were the first mammals domesticated for research purposes, and rats in the laboratory may well have saved as many human lives through the years as they have taken.

Today, there are 51 species within the genus *Rattus*. Norway rats originated on the plains of Asia (northern China), while black rats originated further south in the Indo-Malayan region. Both traveled to Europe with humans although Norway rats came somewhat later. The first reliable accounts of the presence of the more aggressive and larger Norway rats in Europe date back to the 18th century and it was during this time that they began to displace black rats all over Europe. Today, Norway rats have almost completely replaced black rats in Europe and America, where black rats are now rare or absent. In contrast, in tropical zones black rats are more common.

A rat is an amazing creature; it can collapse its skeleton which allows it to wriggle through a hole as narrow as three-quarters of an inch. An adult's jaws are hundreds of times more powerful than a human's. Rats can gnaw through bone, wood, iron or concrete. They also can alert family members by making high pitched sounds to alert them of danger or presence of food. Rats are nighttime animals and detect motion thirty feet away even in nearly total darkness. For locations in the dark, rats also rely on the long whiskers on their faces and hair on their coats to feel their way. As acrobats, rats can scale a brick wall straight up and survive a fall from a five story building. They are skilled climbers and find it easy to board ships by ascending mooring ropes. Rats are indefatigable, and have been known to run five to ten miles a night in wheel cages.

Fortunately it is possible to kill rats by poisoning them because they do not have the ability to vomit or burp. Rats, however, have an acute sense of smell and taste and have learned to scrupulously avoid food that makes them sick.

Rats are vectors and carry at least eighteen diseases that affect people. The most serious are rat bite fever (streptobacillary and spirillary) , Lassa fever, typhus, poliomyelitis, meningitis, trichinosis, *Salmonella* poisoning, hanta virus, leptospirosis (infectious jaundice) and, of course, plague. It is for the latter that rats continue to carry the heinous reputation as they were at the scene of a number of humanity's greatest calamities. Rats and other rodents are the natural reservoir of plague, a disease that during the middle ages killed as much as one third of the population of most towns and villages.

Rats can learn to crave the same drugs as humans do, including alcohol, cocaine, nicotine and amphetamine and have been known to overindulge. Rats are the preferred model for the study of genetics of many complex diseases including hypertension, non-insulin dependent diabetes, renal disease, autoimmune and behavioral disorders. Rats and humans absorb and eliminate a number of drugs at a similar rate. Their findings suggest that testing new drugs on rats and perhaps one other species of animal may satisfy FDA requirements for preclinical studies The basic internal structures of rats and humans are similar despite the disparities in length and diameter. Similarities extend far beyond gross anatomy. Rats even have personalities, they can be sullen or cheerful depending on their upbringing and circumstances. Surprisingly, rats are generally sociable, curious and love to be touched.

It is unlikely that we will ever be witness to eliminating all species of rats despite what the medical journal advised a century ago. In fact, a recent book about the effects of the earth's climate on extinction predicts that the year 1 million will be witness to islands full of novel tree rats, ground rats, slow rats, vegetarian and predatory rats, even diving rats—a scenario too grim to imagine.

Roundworms

—Successors to the fruit fly?

THERE ARE A number of simple organisms used as models for biological research, perhaps the most unusual and likely the most important is the roundworm, *Caenorhabditis elegans*. Model organisms including fruit flies, mice and zebrafish, are the bedrock on which the tower of basic biology is built and used extensively in genetic research. Such research helps to develop a complete understanding of the genetic control of development and behavior. Dr. Sydney Brenner was one of the first scientists to use the roundworm in his research in biology and neurology. It subsequently became the true life's work for Dr. John Sulston. The organism is a small (about 1 mm in length), non-parasitic, free living, transparent worm (nematode), found in the soil. It feeds on bacteria, and can be easily housed and cultivated in large numbers (10,000 worms per petri dish) in the laboratory. The worm grows from fertilized egg to adulthood in a mere three days, and has a life span around two to three weeks under suitable living conditions. Because the worm is transparent, it can easily be observed. *C. elegans* is multicellular and develops from a fertilized egg to an adult just as a human being does. According to Natalie Angier in her book, the Beauty and the Beasty, "C.

elegans is a proxy for the rest of us, a specimen that can be manipulated, irradiated, assortatively mated, plucked apart, scrambled up, over-easied, put back together, sacrificed, and finally understood in a way that a human being could never be."

Dr. Sulston was renowned for his work ethic and happiest working in the laboratory and while there developed a number of freezing, drying and handling techniques. His findings on genetics, together with Sydney Brenner and H. Robert Horvitz, won each of them the Nobel Prize for physiology in 2002 and helped prepare the scientific world for the project of mapping the human genome. Dr. Sulston described their work modestly, as the ability to read the language of evolution. To fully understand the significance it is necessary to know more about the makeup of the human genome, which makes up a complete set of human genes, and comes packaged in twenty three separate pairs of chromosomes. Of these, twenty two pairs are numbered in approximate order of size, from the largest (number1), to the smallest (number 22), while the remaining pair consists of sex chromosomes: two large X chromosomes in women, one X and one small Y in men. One set of the genome comes from the mother and one from the father. Each set includes the same 30 to 80 thousand genes on the same twenty three chromosomes. (The chromosomes are located in the nucleus of each of the 100 trillion cells found in the human body. Inside the nucleus are two complete sets of the human genome, except in egg cells and sperm cells, which have one copy each, and red blood cells which have none.)

The research performed by Drs. Sulston, Brenner and Horvitz, was monumental and instrumental in discovering the genetic coding needed to make a human being. Surely this may be the most celebrated scientific triumph of the 20th century, offering treatments for a whole host of neurological and other diseases. One example is human leukemia where large numbers of immature white blood cells, which normally die before entering the blood stream, are found in the patient's circulation. The study of programmed cell death in *C. elegans* helps to understand why the same process does not occur in human patients.

Dr. Sulston was the work horse of the group, driven and known for his tenacity, for years he sat in a darkened room peering through a microscope in the Medical Research Council's Laboratory of Molecular Biology up to twelve hours per day watching cells dividing, first in the worm larva and then in its embryo, day after day, until he had the entire lineage from one to all of the 959 cells that made up the organism. Of these, 302 are nerve cells, and Dr. Sulston was able to determine the complete wiring diagram, and know what every neuron looks like, how it branches, and how the connections are made with other nerve cells. Some of the cells mysteriously died within 30 minutes paving the way for the surviving cells to fulfill their function. This work has been shared via the internet and with other researchers from around the world. The collaboration has provided a wealth of information and the means to understand the massive cell degeneration found in Alzheimer's, Parkinson's and other diseases of aging.

Salt

—Widely used but often misunderstood.

SALT (SODIUM CHLORIDE) has been used for millennia and yet most people have no idea about its effect on health or its history. If you are confused—join the crowd. Sodium, which the body cannot manufacture, is an essential element needed for proper fluid balance, nerve transmission and muscle contraction, and it along with potassium, calcium and magnesium play a vital role in regulating blood pressure. There have been a number of claims that reducing dietary sodium (which represents 40 percent of the salt molecule) is crucial to our well being and some that report that doing so is a health hazard. The majority of scientific findings, however, suggest that most Americans should cut back on sodium. Excess sodium is responsible for most cases of high blood pressure which is the leading risk factor for heart attacks, strokes and kidney failure. High blood pressure continues to be a health problem: today one in three adults is afflicted, and of those only half have it under control. Another third of adults have blood pressure higher than normal, though not yet in the high blood pressure range. Poor treatment remains a serious contributor to heart disease and deaths.

Many factors are responsible for the excess of salt in our diet. Unfortunately, there is an innate response that drives a human to seek and ingest salt containing foods and fluids. This means that for many of us our daily salt intake is in excess of physiological requirements. A major reason is that the majority of salt is added to our foods by food and drink processors and restaurants, and not from our salt shakers.

The recommended daily intake for healthy American adults is only 2300 milligrams of sodium per day, or the amount in about 1 1/8 teaspoons of salt. This information will be provided in the new nutrition facts labeling, scheduled to take effect beginning in mid-2018 until January 2021. Currently the average American consumes more than 3400 mg per day, an amount often found in a single restaurant meal. According to a report in the New England Journal of Medicine, an average reduction of just 400 milligrams of sodium a day could save 28,000 lives and $7 billion in health care costs yearly. Cutting back, therefore, should be fairly simple, avoid processed meats, buy low sodium or sodium free products like soups and condiments, and use less salt when cooking your own meals at home.

In addition to health matters, there is some interesting history surrounding salt. Until modern times it provided the principal way to store food and Egyptians used it to make mummies. Salt has the ability to preserve, protect against decay, and to sustain life. On every continent, once human beings began cultivating crops, they began looking for salt to add to their diet. How they learned of this need is a mystery. Another development that created the need for salt began when animals were raised for meat rather than killing wild ones. Animals, like humans, need salt.

Almost no place on earth is without salt, and so for all of history until the twentieth century, salt was desperately searched for, traded for, and fought over. For thousands of years, salt represented wealth. Many governments taxed it to raise money for wars. Soldiers and sometimes workers were paid in salt. (The Latin word for "salt" was sal, and the "salt money" given to soldiers was called salarium. Salarium later became used for the regular pension or salary paid to soldiers.) An excellent book entitled "Salt—A World History" written by Mark Kurlansky provides a remarkable story about salt and how it influenced the establishment of trade routes and cities, provoked and financed wars, secured empires and inspired revolutions.

While the historical aspects of salt make interesting reading, its use as part of our well being is much more important. Reducing the intake of salt appears to be a prudent dietary consideration in the management of high pressure. Few risk factors are as important to health. High blood pressure is second only to smoking as a preventable cause of heart attacks and strokes and heart disease remains the leading killer of Americans. Under new guidelines high blood pressure will be defined as 130/80 millimeters of mercury or greater for anyone with a significant risk of heart attacks or stroke. The previous guidelines defined high blood pressure as 140/90. (The first number describes the pressure on blood vessels when the heart contracts, and the second refers to pressure when the heart relaxes between beats.) To calculate your risk of heart trouble, contact your physician and use the online calculator sponsored by the American Heart Association. The calculator will ask for your blood pressure and cholesterol measurements. Try ccccalculator.ccctracker.com, it can help to safeguard your health.

Scorpions

―――᭜―――

—Scary and nasty creatures!

ACCORDING TO NATALIE Angier, in her book *The Beauty of the Beastly*, scorpions deserve multiple entries in the *Guinness Book of Records*; they are some of the biggest, meanest, longest lived, most sensitive, most maternal, least fraternal, slowest, quickest, and most luminous creatures among the arachnids and insects. The latter feature is the ability to glow under ultraviolet light. Their exoskeleton is made of a tough layer of tissue composed of a cuticle protein, chitin. This coat reflects the ultraviolet rays from moonlight and other light sources so brightly that even a black scorpion will be a fluorescent shade of green or pink. Fossilized scorpions from 300 million years ago still gleam brilliantly with exposure to ultraviolet light. The glow may have evolved to attract insects. Scorpions are equipped with a venom contained in a gland on the back of its tail, which the animal can whip forward in a fraction of a second to sting a victim, sometimes repeatedly. The venom is comprised of up to thirty neurotoxins, each designed to kill a different kind of prey, some against insects, others are best at paralyzing frogs, lizards or other small vertebrates. Scorpions, like scores of other species, evolved venom as a tool for defense and predation. Often, scorpion's venom will develop in response to specific adaptations in their predators or prey, triggering a spiraling co-evolutionary process that produces more lethal venom in the scorpions and more robust resistances in their enemies. Scorpions have eight legs and two body segments, including the cephalothorax and the abdomen. They develop through successive stages starting with eggs which hatch miniature versions of the adults that molt and grow in size. Their speed of growth and development varies. Assuming scorpions avoid being consumed by owls, bats, snakes and other animals, they have the potential to live fifteen to twenty years, and beyond, longer than any other known arachnid or insect. Contributing to the scorpion's longevity is its exceeding low metabolic rate, which is slower than that of any other invertebrate. Creatures with slow metabolisms generally live far longer than those which burn energy at a rapid clip.

Scorpions are also one of the most feared arachnids and closely related to spiders, mites and ticks. They can be found at every corner of the globe and on six of the seven continents, from the southern tip of South America to the arid expanse of the Sahara desert. Scorpions are hardy and have been around for hundreds of millions of years. There is reason to be afraid of scorpions, their venom can kill you or send you to the emergency room. However, there are more than 2,200 species of scorpion, and only about thirty of them might hospitalize you and, of those, symptoms vary based on age and health of the victim. Most of the really harmful scorpions, the types that induce symptoms that could be fatal if not properly treated, are distributed throughout Asia, the Middle East, some in Africa, but very few in the United States and quite a few in South America, particularly Brazil. Surprisingly, the venom of one species may be used for a medical advantage.

Imagine an imaging agent that "lights up" or stains malignant tumors and other cancers. This "tumor paint" is derived from an Iranian species commonly called the deathstalker scorpion. It produces a toxin that targets certain brain tumors to stop their growth. The venom could potentially help surgeons resect tumors with the least amount of extraneous damage to surrounding noncancerous tissues. Tumor painting could thus provide a new pathway for both the diagnosis and treatment of tumors. Tumor paint is currently being studied in gliomas, which are among the deadliest forms of cancer and are the leading cause of primary brain tumor growth. In brain tumor resection surgery, even small errors during excision can have devastating neurological consequences Thus, newer technologies are needed that can accurately distinguish between the tumor and normal brain tissue and guide tumor resection in real time. The deathstalker's venom also contains a small protein that can be used to carry drugs across the blood brain barrier. (The barrier serves to protect the brain from toxic substances and prevents many drugs from being used.) This finding may allow life saving drugs to be used to treat neurological diseases and tumors.

Recent evidence has availed the fact that some varieties of desert scorpions are master architects and builders. Burrows used by these species for warmth to increase their body temperature have been found to follow a very sophisticated design beginning with a short vertical entrance shaft that flattened out a few centimeters below the surface into a horizontal platform. The burrows then turn sharply downwards, descending further below the ground to form a dead end chamber. This cool, humid chamber where evaporation water loss is minimal, provides a refuge for the scorpions to rest during the heat of the day.

There is still much to be learned about the remarkable abilities of these frightening, carnivorous creatures.

Sleep

—*Still a mystery!*

Most of us will spend a full third of our lives asleep, and yet the majority of us do not have the faintest idea of what sleep does to our bodies and our brains. Sleep is still one of secrets of science. Combine that with the fact that sleep deprivation or insomnia has become a major health problem in the United States that crosses all economic lines. A recent report indicated that nearly one third of all working adults get six or fewer hours of sleep each night as opposed to the recommended seven or eight. Insomnia is quite prevalent and affects a large share of the population. This is a serious issue and associated with an increased risk of mortality in men. Recent studies suggest that insufficient or disturbed sleep is associated with metabolic disorders such as type 2 diabetes and obesity. There is also ample evidence indicating that difficulty in falling asleep and non-restorative sleep are associated with the risk of cardiovascular (heart) disease in adults and in young children. There is much still to be learned about the cause, stages, benefits, and insomnia, the most common sleep disorder.

According to a recent study, the body uses two mechanisms to regulate sleep. One is the body clock, which attunes humans and animals to the 24 hour cycle of day and night. The other mechanism is the sleep ""homeostat" , a device in the brain that keeps track of your waking hours and puts people to sleep when resetting is necessary. This mechanism represents an internal nodding off point separate from external factors.

There are two kinds of sleep—REM (rapid eye movement) sleep and non-REM (NREM) sleep. Rapid eye movement is the stage where the eyes dart quickly back and forth under the eyelids. Dreaming occurs during this stage. The brain is active during REM sleep and brain waves resemble awake patterns. Breathing is rapid, shallow and irregular, and heart rate and blood pressure increase. This stage plays an essential, but not fully understood function. REM sleep stimulates areas of the brain responsible for learning and memories. There are three stages on NREM sleep. These range from drifting in and out of sleep (stage N1), through light sleep (N2) and into deeper sleep (N3). Stage N2 accounts for 40 to 50 percent of sleep time. Stage N3 is the physically restorative, as brain waves slow considerably. Older adults experience less N3 sleep than younger people and suffer a reduction in sleep efficiency (the percent of time of actual sleep). The lower the sleep efficiency score, the higher the mortality risk, the worse the physical health, and the lower the cognitive function, typified by forgetfulness.

According to recent research, sleep is important because it weakens the connections among brain cells to save energy, avoids cellular stress and maintains the ability of neurons to respond selectively to stimuli. The scientist conducting the study noted that sleep is the price we pay for learning and memory. While we are awake, learning strengthens the synaptic connections throughout the brain, increasing the need for energy and saturating the brain with new information. Sleep allows the brain to reset, helping integrate newly learned material with consolidated memories. Studies in mice also indicated that the channels in certain cells of the brain can help remove a toxic protein called beta amyloid from brain tissues during sleep. Beta amyloid is known to accumulate in the brains of patients with Alzheimer's disease. Unfortunately, while there are a number of theories as to the purpose for sleep there is still no agreed upon definition of what sleep actually represents. This is further confounded by the fact that some species require a 19 hour period of inactivity while others only two.

Insomnia is defined as difficulty with the initiation, maintenance, duration, or quality of sleep that results in the impairment of daytime functioning, despite adequate opportunity and circumstances for sleep. Transient insomnia lasts less than a week, and short term insomnia one to four weeks. Chronic insomnia—insomnia lasting more than a month—has a prevalence of 10 to 15 percent and occurs more frequently in women, older adults, and patients with chronic medical and psychiatric disorders. The Centers for Disease Control and Prevention estimate that about 70 million Americans suffer from sleep problems.

As explained above, there is still much to be learned about the science of sleep. But until that mystery is solved, those of us with poor sleep habits you would be wise to do the following: (1) keep a regular sleep schedule, (2) avoid stimulating activities within 2 hours of bedtime, (3) avoid caffeine, alcohol and nicotine in the evening, (4) avoid going to bed on a full or empty stomach; and (5) to sleep in a dark, quiet, well ventilated space with a comfortable temperature.

Submitted by Max Sherman.

THE SENSE OF SMELL

—Dogs or us?

THERE WAS A recent report that researchers in France are working on a sensor to detect explosives carried by individuals who hope to smuggle them through airport security systems. The hope is to create a device that could supplement or even supplant the best mobile bomb detector—a sniffer dog. Dogs have a scent sensitivity many thousands of times greater than humans. In fact, this ability is one of the more curious facts in nature. Humans, however, and other primates have relatively good senses of smell. This is true even though human evolution has been characterized with a gradual increase in vision and a reduction in the ability to smell. The change is likely due to a progressive diminution of the nose over many eons as the eyes gradually moved to the middle of the face to enhance the depth of vision. There is little doubt that over many years this facial change has resulted in a progressive reduction in the proportion of functional olfactory receptor genes. Mice, for example, have approximately 1300 olfactory receptor genes, of which some 1,100 are functional, whereas, humans have only some 350 functional genes (smell receptors) of approximately 1,000. (Scientists still cannot explain how the 350 smell receptors are able to account for the human ability to detect thousands of different odors.) Humans, however, perform as well as, or better than other mammals. Many humans actually make a living with their noses, think of oenologists (wine experts), perfumers, and food tasters. The subject matter may appeal to those of us with a nose for news about the nose.

Smell is the most direct of all senses. Only substances volatile enough to spray microscopic particles into the air have an odor. (This ability allows us to enjoy the taste of food.) The nose, where the ability to smell begins, is composed of two nasal cavities separated by a middle wall, the nasal septum. Initial detection of odors takes place at the posterior of the nose, in the small region known as the olfactory epithelium. Odor molecules flow back into the nasal cavity behind the bridge of the nose where they are absorbed by the mucosa containing receptor cells bearing microscopic hairs (cilia). Specialized proteins, known as receptor proteins extend from the cilia. There are approximately five million of these cells which then send impulses to the brain's olfactory bulb or smell center via fibers known as axons. The olfactory epithelium also contains neuronal stem cells which generate olfactory neurons though out the life span. Each of these neurons express only one odorant receptor. The receptor is activated by a specific odor which in turn activates a specific region in the brain. Unlike most neurons, which die and are never replaced, the olfactory sensory neurons are continually regenerated.

Sniffer dogs

Scientists have known since the 1950s that dogs and other keen scented mammals such as rats and rabbits have a specialized anatomical structure in their nasal cavities. Called the olfactory process, it is a large maze of highly convoluted airways that humans and other primates lack. In dogs, the recess lies right behind the eyes and takes up almost half of the interior of the nose. Using a computer model of the canine nose, scientists have discovered that when a dog sniffs, each nostril pulls in a separate odor sample. The dog can tell which nostril is pulling in the scent so it knows which direction to go when tracking. The dog's nose has a unique nasal airflow pattern, which helps transport odor molecules quickly via a single airway to the olfactory recess. There are several differences between the human and dog olfactory systems, which explains the dissimilarity in olfactory acuity or the sensitivity of the sense of smell. The first is the size of the nasal cavity and the amount of air that can be inhaled. German shepherds, for example, can breathe in five times more air than a human. Dogs have more olfactory sensory cells than humans; estimates of 5 million in a human compared to 125 million in a dachshund and 300 million in the bloodhound. The dog's olfactory bulb is about 40 times larger than in the human. These differences result in a human being able to smell a mixture of odors compared to the dog that can smell a range of distinct and different scents. The dog also has something else a human doesn't; the vomeronasal gland. The gland is actually a pair of long, fluid filled sacs that opens into either the mouth or the nasal cavity and located in the area above the roof the mouth. It appears to be an accessory olfactory organ that allows dogs to identify scent and possibly pheromones.

In a recent book about dogs written by Alexandra Horowitz, she summarized the difference between dogs and humans, thusly: "What dogs see and know comes through the nose, and the information the dog has about the world based on smell is unthinkably rich. It is rich in a way we humans once knew about, once acted on, but have lost, and have since neglected."

How important it is to hear!

―――∞―――

One of my favorite quotations regarding the abstract nature of sound is as follows:

" *The brain is composed of 100 billion electrically active cells called neurons, each connected to thousands of its neighbors. Each neuron relays information in the form of miniature voltage spikes, which are converted into chemical signals that bridge the gap to other neurons. Most neurons send these signals many times per second; if each signaling event were to make a sound as loud as a pin dropping, the cacophony from a single human head would blow out all of the windows.*"

This quotation, in addition to being extremely well written, relates to the incessant action occurring within us that we know little about. Hearing, however, is an extraordinary sense despite its inability to detect the sound of the routine continual activities of the brain or other body organs. Audible sound constantly surrounds us and informs us about many objects in our world. And our ability to determine the sources of sounds is one of our most important biological traits. Unfortunately, hearing loss is the most common sensory defect in humans, affecting normal communication in more than fifty percent of people aged 65 or older. (For those older than 85, it is eighty percent.) Half a billion people, almost seven percent of the global population, had disabling hearing loss in 2015. Thus, the mechanism of hearing loss is a subject we aging individuals should be acquainted with.

The auditory system is composed of three anatomical compartments: the outer, middle and inner ear. Sound waves impinging on the head are captured by the outer ear and conveyed through the external auditory canal to the tympanic membrane. Vibrations of the tympanic membrane, caused by airborne sound waves, are transmitted through the middle ear to the inner ear by a chain of movable bones. These bones, or ossicles, consist of the malleus, which is connected to the tympanic membrane; the stapes (the smallest bone in the body), which is attached at its base to the oval window of the vestibule; and the incus, which is situated between the malleus and stapes and articulates with both. The tympanic membrane is held in place by fibers and cartilage situated in a bony groove between the outer and middle ear. The major function of the middle ear is to provide an effective and efficient means to deliver sound to the inner ear, where the neural process of hearing begins.

The inner ear consists of the bony and the membranous labyrinths that are filled with fluid. One part of the middle ear, the cochlea, processes auditory signals, whereas our sense of equilibrium depends on the vestibular apparatus composed on the three semicircular canals. The inner ear is an evolutionary triumph of miniaturization, owing its success to its complement of hair cells that are responsible in the cochlea for our sensitivity to sound.

Evolutionary success of hair cells is evidenced by the persistence of structurally similar inner ear anatomy in all vertebrates. The 32,000 hair cells of the two cochleae and the total of 134,000 cells found in the organs of equilibrium forward their signals to the complexes of the brainstem's medulla and pons. Destruction of hair cells results in loss of equilibrium and profound deafness.

Sound may be defined in physical terms, it requires vibration of an object . Any object with the properties of inertia and elasticity may vibrate, and produce sound. If we hear the vibration, the sound is audible. A vibrating object causes a wave motion in air, which then causes the eardrum to vibrate, starting the process of hearing. Sound can travel through any elastic medium that has inertia (a force that is exerted on an object to make it move). Thus sound can travel through air, water, steel, etc, but not through a vacuum. Once a sound's pressure wave has traveled from its vibrating source, it will eventually encounter the outer ear of a person and the process of hearing begins. Surprisingly, children and adults depend more on their right ear than the left for processing and retaining what they hear.

There are a number of conditions that lead to hearing loss. The most common cause in childhood is otitis media with effusion, in which the middle ear and mastoid air spaces are filled with serous fluid. Other causes include genetic mutations, noise exposure, drugs, smoking and diabetes.

At present, 50 million Americans suffer some degree of hearing loss—17 percent of the population. And hearing loss is not exclusively a product of growing old. The usual onset is between the ages of nineteen and forty-four, and in many cases the cause is unknown. Fortunately, hearing loss can be treated, but not cured, through hearing aids or cochlear implants. There is hope, however, research in gene therapy, molecular therapy, stem cell technology, and the regeneration of hair cells is advancing at a breathtaking rate.

Spanish Flu

—*May it never return.*

Health officials continue to urge all Americans 6 months and older to receive an annual flu shot—except of course for those who have ever had a severe or life-threatening allergic reaction to the vaccine. Influenza is nothing to fool with especially for pregnant women, the very young, the elderly, or any whose immune systems are suppressed. Flu is one of the most common infectious airborne (spread by sneezing, coughing or talking) viral diseases that occurs in seasonal epidemics and can cause variable degrees of symptoms, ranging from fatigue to respiratory failure, including red and watery eyes, high fever, sore throat, headache, nasal discharge, weakness, cough, bloody sputum, and, in the worst case, mortality when bacteria swarm into the injured lungs causing pneumonia. The Centers for Disease Control and Prevention estimates that seasonal influenza is responsible for more than 20,000 deaths annually.

While 20,000 deaths is something to be alarmed about, it pales in comparison with those who died in the great Spanish influenza pandemic of 1918-19. (It was called "Spanish" flu because more than 8 million Spaniards were ill including King Alfonso XIII, and the cases were widely reported.) This was the disease responsible for the deaths of approximately 50 to 100 million people world wide, many between 20 and 40 years of age, and it stands out as the deadliest single event in recorded human history. The disease has plagued mankind through centuries and the word "influenza" started being used towards the end of the middle ages.

Most everyone living in 1918 suffered from the flu, it killed 2.5 percent of its victims, and a fifth of the world population suffered from it. So many died, that the average life span in the United States fell by more than twelve years that year. In comparison, AIDS killed 11.7 million through 1997, World War I was responsible for 9.2 million combat deaths and around 15 million total deaths. World War II accounted for 15.9 million combat deaths. The cause of the pandemic and the reasons for its severity had remained one of the most discussed medical mysteries throughout most of the 20th century. Influenza viruses were finally isolated about 15 years after the pandemic, but scientists of the early 20th century were then not capable of understanding the emergence, pathology (aspects of the disease), or disappearance, let alone determining a means of prevention. Many years past until 1996, when with new techniques it became possible to recover and sequence fragments of the viral ribonucleic acid (RNA) retained in preserved tissues from several 1918 victims. The full genome sequence is now available and this information may provide the ability to prevent and control future pandemics. Of note is that several different strains of pandemic viruses (1946, 1956, 1968 and 2009) since 1918 contain gene segments derived from 1918 virus but none approach its lethality.

Molecular biologists know that the simple influenza virus has only eight genes, each made of ribonucleic acid (RNA), and that the viruses die in hours if left alone with no cells to infect. They even know what the flu viruses look like under an electron microscope—- they are egg shaped particles, which sometimes form long filaments. The flu virus particles are wrapped in a slippery fatty membrane, held in place by a protein scaffold. Biologists also understand how viruses burrow into a human cell and burst out again by using hundreds of sharp edges that poke out of the virus's membrane. They even know why human influenza viruses infect only cells of the lungs, the only human cells with an enzyme the virus needs to split one of its proteins during the manufacture of new virus particles. Flu viruses depend upon two sorts of proteins to enter and exit the cells; one, hemagglutinin, is used to hoist itself into a cell, and the other, known as neuraminidase, is used by newly made viruses to burst the cell open so they can escape in a spray and infect new cells. The hemagglutinin and neuraminidase proteins also define the flu strain, both the 1918 and 1946 strains were H1N1, the next genetic change was in 1956, with strain H2N2. The one that arrived in 1968 involved a virus whose hemagglutinin had changed from the 1956 virus but whose neuraminidase had not, it was thus named H3N2. Even more confusing, is that there are actually four types of influenza viruses—- A, B, C, and D. Influenza A (H1N1) and H3N2) and one or two Influenza B viruses are included in each year's vaccine. Vaccines are updated annually to keep up with changing viruses. But it can take days for the body to develop enough antibodies to stop a flu infection, unless that flu strain has invaded the body before. In that case, the immune system can quickly marshal its forces and block the virus before the infection occurs. The annual recommended flu shot contains the strains most likely to occur and the reason why each of us should be vaccinated. Last year's flu vaccine, however, was only 20 to 30 percent effective due to a mutation in the H3N2 strain. No matter, it is better to get a flu shot than suffer from the flu.

Spiders

—*Vile but valuable!*

ARACHNOPHOBIA (AFRAID OF spiders) is quite common and with good reason—such fear is deeply embedded in us and found to be hereditary. Arachnophobia has a long history, as early as the time of the birth of Christ parts of Abyssinia (officially Ethiopia) were abandoned by the whole population as a result of a plague of spiders. Spiders rank with snakes, cockroaches, and rats on the list of most disliked creatures. According to some people, however, spiders are fascinating and often attractive animals, and like rats may have a purpose in the medical field.

Spiders are the primary predators of insects, possess a venom system to assist in capturing prey, and produce and utilize silk in many more ways than any other animal. About 13,000 of the known species manufacture webs . Anatomically, spiders are animals without backbones (invertebrates). They have two parts to their bodies, a cephalothorax (the head and thorax are fused together) and an abdomen. Eight legs and usually eight simple eyes are on the cephalothorax. Most spiders have teeth to chew an insect's hard exoskeleton. Spiders then expel juices that liquefy their prey's insides and allow them to be swallowed.

The evolution of myriad spider silks is reflected in the dazzling abundance of web types. All designs derive from a simple silk mesh used by ancestral spiders to line earth burrows more than 380 million years ago. Garden spiders can spin an orb web (a web with a spiral pattern) that is invisible to humans but visible to insects. Ogre faced spiders hang upside down from silk threads attached to branches. There are also social spiders that work together to make extremely large webs that are attached to branches of trees and shrubs. By working together social spiders can catch and eat insects may times larger than they are. Strangely, spiders do not become attached to their webs. Their legs feature a disengaging mechanism that enables the arachnid to detach itself instantly from a sticky strand.

Spiders are legendary as the materials-science experts of the animal kingdom. They can produce as many as seven different kinds of silk. Silk from spiders is a protein fiber that has been the subject of intense research because of its impressive mechanical properties, including high strength and unprecedented toughness. The toughness of silk fibers is superior to any of the best synthetic high performance fibers available today.

To begin a web, a spider anchors a strand of dragline silk—three times stronger than the Kevlar used in bullet proof vests—and waits for a breeze to blow it to a second attachment point. The spider then completes the outer ring and spokes, and finally builds a spiral. Dragline silk is several times stronger than steel, on a weight by weight basis, but a spider's dragline is only about one tenth the diameter of a human hair.

Dragline silk is a composite material comprised of two different proteins, each containing three types of regions with distinct properties. One of these forms an amorphous (noncrystalline) matrix that is stretchable, giving silk elasticity. When an insect strikes the web, the stretching of the matrix enables the web to absorb the kinetic energy of the insect's flight. Spiders are able to produce several mechanically distinct fibers from different silk glands. These different silks include dragline silk, which is stiff and strong, and capture silk, which acts like a very stretchy rubber. The capture spiral in an orb web is stretchy and can triple in length before breaking. Spiders use silk for a number of activities central to their survival and reproduction, including wrapping of egg sacks, preparing safety lines, lining retreats, and most famously to capture insects.

Spider webs have been used as dressings for wounds and even as fishing nets, but the silk itself has found employment only as cross hairs in optical instruments. This historical lack of use is about to change. As mentioned above, spider silk has the unusual combination of high strength and extensibility, characteristics unavailable in synthetic materials. Now that there a greater understanding of the protein composition, the spinning process, chemistry and physical properties, and the genetic sequence, it is more likely that silks will be used in a number of applications.

Molecular biologists are planning to use the proteins from super strong dragline silk to build artificial tendons and ligaments and in improving the quality of microphones in hearing aids. The researchers needed more silk than they could harvest from spiders in captivity, so they genetically engineered goats to produce the proteins in their milk. After the silk proteins are extracted and purified, a machine spins them into the needed fibers. Researchers at Notre Dame have genetically engineered silkworms to produce spider webs, a process that began by inserting spider DNA into ordinary silkworms. To most of us spiders are vile creatures and we inherently fear them. Scientists on the other hand are just beginning to discover how valuable spiders can be.

Stanley Cohen

—He made a breakthrough in cellular growth!

As old as I am, I continue to be amazed and impressed by individuals who come from a humble background and yet manage to make a significant difference in our lives. One such person was Stanley Cohen. He was one of four children of a poor and immigrant tailor father and homemaker mother from Russia, born on November 17, 1922 in Brooklyn, New York. As a child he contracted poliomyelitis, which left him with a persistent limp—not a propitious, memorable or encouraging beginning to someone's life. His infirmity did not hamper his curiosity or search for knowledge—he had an interest in embryology and learning how things work by taking them apart. Cohen began his education as a biology major at Brooklyn College in 1945, while there he became fascinated by embryology (the study of unborn children or animals). Of particular, his desire was to know more about the chemical reactions involved inside the egg or embryo. For this reason he obtained a double major in chemistry and biology. After graduating he worked as a graduate assistant at Oberlin College. He continued his education towards a Ph.D. degree at the University of Michigan. While there he studied the Krebs cycle in common earthworms that he collected from the university campus. (The Krebs cycle is the sequence of action by which living cells generate energy.) His dissertation was based on the metabolic function of the earthworm. His first real job as he called it, was at the Pediatrics Department at the University of Colorado where he was granted a fellowship.

Dr. Cohen's true love materialized later at Washington University where he continued to pursue studies in embryology and biochemistry. He joined Dr. Rita Levi-Montalcini and both were instrumental in isolating a nerve growth factor (NGF), a factor Levi-Montalcini had found. There he identified another cell growth factor in the chemical abstracts containing NGF. Cohen discovered that this substance caused the eyes of new born mice to open and their teeth to erupt several days earlier than normal. He called this substance "epidermal growth factor" or EGF and eventually purified it and analyzed its chemical structure. Dr. Cohen found that EGF influences a great range of developmental events in the body. He also developed the mechanism by which EGF is taken into individual cells and acts on them. This involved identifying receptors on the cell surface on which the growth factor acts that explains how such proteins change the biology and behavior of individual cells.

Dr. Cohen eventually moved to Vanderbilt where he resumed his research in growth factors and was promoted to full professor of biochemistry in 1967. In 1976, he became an American Cancer Society research professor of biochemistry. Dr. Cohen died on February 7, 2020 in Nashville, TN, he was 97.

Drs. Cohen and Levi-Montalcini helped to unravel the basic mechanisms underlying cell growth and the means in which developing cells connect to others. They shared the Nobel Prize in medicine and physiology for their research. These discoveries paved the way for finding treatments for a number of cancers such as lung, breast, head and neck and other gastrointestinal varieties. Targeting the EGF receptor has led to dramatic tumor responses and often rescues the patient from near death. Few discoveries have done as much to transform clinical care. In addition to cancer, Dr. Cohen's discoveries have led to current clinical trials where EGF families of growth factors are being used as treatments for heart failure. Growth factors in the EGF family are now recognized as clinical regulators of not only cardiac development but also the maintenance of the adult heart. And there is increasing evidence in animal studies that the EGF receptor pathway may be involved in progressive kidney diseases including those associated with high blood pressure and diabetes. Because EGF influences a number of developmental events in the body it is being used to stimulate nerve cell growth, wound healing and cornea repair. It may also improve the effectiveness of skin transplantation. Since the discoveries of NGF and EGF, thousands of related papers and numerous reviews have been published revealing new aspects of growth regulation. Dr. Cohen's work has led to discoveries of receptors of other growth factors and to eventual more targeted clinical therapy. Growth factor research has provided a deeper understanding of other medical problems like deformities and senile dementia.

FINAL THOUGHTS

In addition to the Nobel Prize, Cohen was awarded many honors and medals throughout the world. He was also presented with the National Medal of Science. His fame was worldwide, including being honored by a stamp issued by Uganda.

Stanley Falkow

—*Microbiologist extraordinary.*

To the average reader it might appear that my columns generally feature someone prominent in science or medicine, and I do tend to focus on those I deem unsung or forgotten heroes. Among them I have chosen to include Otto Warburg, Emory Rovenstine, John Sulston, James Parkinson and David Nachmansohn. There , of course, have been many others, and one, Stanley Falkow, just passed away this year. He died on May 5 of complications from myelodysplastic syndrome. Dr. Falkow has been called the Father of Molecular Pathogenesis, a title earned because of his outstanding work on bacterial resistance to antibiotics and how bacteria cause disease. Bacterial resistance is the ability of bacteria to resist the effects of an antibiotic, it occurs when bacteria change in a way that reduces the effectiveness of drugs, chemicals, or other agents designed to cure or prevent infections. The bacteria survive and continue to multiply, causing even more harm. According to the Centers of Disease Control and Prevention, antibiotic resistance is one of the world's most pressing public health problems. Antibiotic resistance can cause illnesses that were once easily treatable with antibiotics to become dangerous infections, prolonging suffering for children and adults. Antibiotic-resistant bacteria can spread to family members, schoolmates, and co-workers, and may threaten the community. Such bacteria are more difficult to kill and more expensive to treat. (One example is Methicillin Resistant *Staphylococcus aureus* or MRSA.) Unfortunately, antibiotic-resistant infections are a major health risk estimating to cause approximately 700,000 deaths each year. This number could rise by an additional 10 million annually by 2050.

STANLEY FALKOW

Falkow was born in Albany, NY in 1934. His father born in Kiev, was a shoe salesman, and his mother ran a shop selling corsets. The family moved to Newport, RI when was nine years old. His interest in science was attributed to reading *Microbe Hunters*, a book that featured Louis Pasteur and Robert Koch, world renowned bacteriologists. He earned an undergraduate degree at the University of Maine. After working as a hospital microbiologist in the 1950s, he joined Brown University graduate school in 1957, and was quickly drawn towards the emerging field of bacterial genetics and earned his doctoral degree. During the 1960s, he worked on several studies demonstrating how bacteria could swap genetic material, molecular structures now referred to as plasmids. (Plasmids are genetic structures in a cell that can reproduce independently of the chromosomes.) After that, he spent time as a researcher at Georgetown University and the University of Washington. During those years he demonstrated how the bacteria that can cause gonorrhea could acquire resistance to antibiotics. He also identified a subtype of *Escherichia coli* responsible for some types of life-threatening diarrhea prevalent in developing countries. Starting in the 1970s, Falkow warned that the use of antibiotics in animal feeds could spur resistance to medications and damage human health. While others fretted about the dangers of bacteria, he pointed out their benefits as well, such as bolstering defenses against diseases. Eradicating a microbe might cure one ailment while opening the way for another. In 1981, he accepted the position of chairman of Stanford's Department of Medical Microbiology. While there he focused on the mechanisms by which organisms cause disease.

In addition to his research contributions, Falkow was known for his dedication to educating future generations of scientists, and contributed to numerous textbooks and public lecture series on microbiology. He was awarded the prestigious National Medal of Science in 2015 and was elected to the British Royal Society, and academy of elite scientists.

Thanks to Dr. Falkow, the fight continues to discover and prevent the mechanisms by which bacteria become resistant to antibiotics. Scientists have learned that there are several pathways to resistance. Some can neutralize an antibiotic and make it harmless. Others have learned how to pump an antibiotic back outside of the bacteria before it can do any harm. Some bacteria can change their outer structure so the antibiotic has no means of attachment to the organism it is designed to kill. After being exposed to antibiotics, there are times where one of the bacteria can survive. If even one organism becomes resistant it can then multiply and replace all of the bacteria there were killed. This means that exposure to antibiotics provides selective pressure making the surviving bacteria more likely to be resistant. As Dr. Falkow demonstrated, bacteria can become resistant through mutation of their genetic material and his work is primarily responsible for teaching healthcare personnel and government officials involved in health and human services about the benefits and risks of using antibiotics.

Strange unexplained diseases

I AM FASCINATED by historical mysterious diseases that appear suddenly and then strangely disappear. According to Wikipedia, they are defined as diseases for which the cause has not yet been identified. Reasons are lack of identification of etiology (cause) include lack of professional interest, difficult access, and lack of resources, in addition to being unknown to medicine.

There are a number of examples, one is "Sweating Disease". It first reared its ugly head in England in the summer 1485 and there were four further outbreaks - in 1508, 1517, 1528 and 1551 - before completely disappearing and never to be seen in that land again. It was highly contagious and decimated settlements around England taking thousands of lives. In fact, towns found themselves fortunate if half the population survived. Although studies have suggested that it was not as lethal as the plague, sweating illness caused shock and horror because of the sudden onset and lethality.

The disease was distinguishable from influenza outbreaks that had affected England in the early fifteenth century and the pandemic that affected Europe in 1510. According to physicians at that time, the symptoms of

Sweating disease included great sweating and stinking; fever; redness of the face and body; continual thirst; breathlessness; back, shoulder, stomach and extremity pain; myalgia; headache; cardiac palpitations; and a desire to sleep. One chronicler recorded that people were throwing off their bedclothes and running through the streets of London seeking relief from fever. Others treated their thirst with cold drinks but there was no treatment for the foul smelling perspiration. Many died, either as soon as the fever began or shortly thereafter. Barely one in one hundred escaped death, and those that survived twenty four hours after the sweating ended, were not free from the disease. They continually relapsed and subsequently perished. What was shocking was the speed at which it killed the victims. An excellent review of the disease is contained in *Sweating Sickness:In a Nutshell*, a book written by Claire Ridgeway.

Another strange disease occurred during World War I. The first patient identified was afflicted with prolonged sleep. At first, the treating physician, Dr. Cruchet, wondered if the symptoms could be the after effects of mustard gas or even a new chemical weapon. Other patients soon followed, and in all, Cruchet saw sixty-four similar cases with no standard diagnosis. What was striking about these patients was the unusual range of symptoms. Some had fever; others did not. Most complained of headache and nausea. Strangest of all was how much these soldiers slept. It would have seemed almost serene at first—blank, expressionless faces, free of terror and pain, calmly asleep in row after row after row. What was frightening about these men was that they would not awaken. It must have felt like being in a room full of the breathing dead. At first, their symptoms vaguely matched any number of diseases, but as they progressed, they did not follow the course of other diseases of the trenches. The soldiers were not comatose; they were simply asleep. Cruchet spent several months studying the sick soldiers, and he prepared a paper on the unusual cases he found coming from the trenches in Verdun. He did not know, could not know, that on the other side of the war another physician was witnessing the same. That physician, in a clinic in Vienna, was preparing a similar paper on the same subject. Though the two men could not coordinate their efforts, nor even contact each other, they were witnessing the same disease. By definition, that made it an epidemic and in fact, specifically described as "the forgotten epidemic." It remains one of medicine's greatest mysteries and brilliantly characterized in Molly Caldwell Crosby's book, *Asleep*.

One other disease appeared to be limited to the rural population of Yugoslavia, Rumania and Bulgaria. It was first reported in the New York Times, on December 4, 1966. The disease affected twenty five percent of the inhabitants and is almost unknown in any other region. It strikes the young, teen agers or even children under ten years of age. As the disease develops, the sufferers become anemic and greatly weakened. In the advanced stage it is often fatal. Evidence gathered seemed to indicate that it was connected in some way with the rivers in the region.

There are certainly many others, the strangest of all may be the Dancing Plague of 1518. The events took place in 16th century Strasbourg, a French city. A woman began to dance fervently and without pause. She carried on for almost six days, unable to stop. When she finally did, the dancing had spread throughout the city. Within a week, nearly 40 people were uncontrollably dancing in the street. By the end of the month as many as 400 individuals were taking part in this remarkable, bizarre outbreak and dozens died from exhaustion or exposure.

Stuart Levy

—A pioneer in the proper use of antibiotics

ANTIBIOTICS BECAME AVAILABLE after World War II, and have saved countless lives by wiping out bacterial infections. Today, however, many antibiotics are becoming much less effective as bacteria have found a way to build up resistance to them. This will have major health consequences. In fact, antibiotic resistance has the potential to affect people at any stage of life, as well as the healthcare, veterinary, and agriculture industries, making it one of the world's most urgent public health problems. Antibiotic resistance initially a problem in hospitals and developing countries, today affects the world at large. Each year in the U.S., more than 2 million people are infected with antibiotic-resistant bacteria, and at least 23,000 people die as a result. The numbers are likely to rise with population growth and the greater incidence of resistance. Fortunately there were scientists who saw this coming. One of them was Dr. Stuart Levy, who may have been the first researcher to predict the occurrence of multidrug resistance and that antibiotic –resistant organisms could leap from animals to humans.

Dr. Levy was a physician at Tufts University, he died at

age 80 after an extended illness. He received his bachelor's degree from Williams College and a medical degree from the University of Pennsylvania. In 1971 he became a professor at Tufts where he conducted research for 47 years until his retirement in 2018. He was the director of the Center for Adaptation Genetics and Drug Resistance at the School of Medicine and president of the International Alliance for the Prudent Use of Antibiotics. He was widely known for his pioneering research. In 1976, his team published a study showing that antibiotic-resistant bacteria could be transferred from the intestinal flora of farm animals to workers, with implications for places such as hospitals and clinics. In the study, chickens were fed antibiotic supplemented feed, and as expected, within one week their intestinal flora contained almost entirely antibiotic resistant organisms. Increased numbers of resistant intestinal bacteria also appeared but more slowly, in farm members, but not their neighbors. The resistant bacteria contained transferable plasmids conferring multiple antibiotic resistances. In 1978, he discovered that one way bacteria resist antibiotics to through efflux pumps that transport proteins that organisms use to extrude substances into their environment. Studying whether antibiotic use in animals was causing problems in humans was a totally novel idea and something that nobody had considered. After Levy published his findings, the scientific community began to investigate antibiotic resistance in other places where antibiotics were frequently used such as hospitals. Now many hospitals have antibiotic or infection control specialists whose job is to collaborate with doctors, nurses and pharmacists to make sure antibiotics are appropriately used. Much of this descended from Levy's influence.

His 1992 book, *The Antibiotic Paradox; How Miracle Drugs are Destroying the Miracle*, was translated into four languages. In addition to this book, his lab group published more than 300 papers. One of those papers, published in the New England Journal of Medicine in 1998, repeated his findings from years earlier. He referenced an incident where a multidrug resistant strain of Salmonella had been associated with hamburger traced to farm using antibiotics to promote growth in cattle. Dr. Levy had urged a re-evaluation of and an eventual discontinuation of the subtherapeutic use of valued human antibiotics, namely penicillin and tetracycline for growth promotion. Little change occurred here, but in Europe and elsewhere, restrictions on the use of antibiotics for growth promotion were legislated in the late 1970s in response to the spread of a multidrug-resistant strain of typhimurium from animals to people. Thus, almost twenty years ago, Dr. Levy forecasted that several strains of disease-causing bacteria in the United States would become untreatable, including vancomycin resistant enterococcus, *Mycobacterium tuberculosis, Pseudomonas aeruginosa, and Acinetobacter baumanii.* The article prescribed a number of suggestions for proper antibiotic use with the warning that given time and drug use, antibiotic resistance will emerge. There are no antibiotics to which resistance has not appeared. An example was penicillin-resistant Streptococcus pneumonia which took 25 years to become a clinical problem. Dr. Levy believed in using antibiotics prudently and with care, not getting rid of them. Today, thanks in part to Dr. Levy, the medical profession has a greater understanding of the advantages and disadvantages of using antibiotics. Despite that, more than 47 million prescriptions issued to patients for antibiotics are unnecessary. Everyone including consumers and healthcare providers should heed the words of Dr. Levy and know that antibiotics must be handled more like a precious resource, than a cure all. The Centers for Disease Control website is a great source of information, see CDC.gov-antibiotic-use.

Final thoughts

Unfortunately, Dr. Levy died last month, but according to his obituary he lived to see the consensus shift from the manner antibiotics are indicated, i.e., when they should be used, for how long, and their administration.

Super Navigators

—Getting around without GPS!

THOSE OF US who rely on the Global Positioning System (GPS) technology, a compass or a map to guide us to our destination must marvel how birds and other animals navigate without similar aid. Incredibly, they find their way using the sun, stars and other landmarks and some can fly up to 56,000 miles a year. According to a new book, entitled *Supernavigators*, ants and bees orient themselves by detecting the polarization of the sun. Bees also look for patterns that they recognize, regardless of whether they are natural or man-made. Dung beetles use moonlight and the Milky Way as their guide. Seemingly delicate, monarch butterflies travel all the way to Texas and Mexico and get there by orienting the sun's position in the sky, or if it is cloudy, with the help of polarized light. The number of animals traveling long distances, from insects to sea turtles, and from eels to whales is astonishing, as are the many means they use to find their way.

David Barrie, the author, tries to address a number of questions regarding the methods animals use to navigate. Examples include how a dung beetle can roll a ball of dung in a straight line or how wasps after flying off on a hunting expedition can find their nest again? What strange sense guides a sea turtle back to the beach where she was born to lay her eggs? And when a pigeon is released hundreds of miles away from its loft, in a place it has never gone before, how can it find its way home? His questions aren't restricted to animals. Even the lowest forms of life, can engage in a surprising means of navigation. The so-called magnetotactic bacteria contain tiny magnetic particles that, when joined end-to-end, act like microscopic compass needles. These "needles" force the bacteria to align themselves with the earth's magnetic field and thereby help them find their way down to the oxygen-poor layers of water and sediment where they flourish. The needles found in bacteria from the northern hemisphere have the opposite polarity to those in the southern hemisphere.

Barrie's book explains the challenge of navigating across a vast ocean without a compass and his book discusses how the animals maintain a steady course using odors, soundscapes and the weak Earth magnetic field. He even tells the story of a pigeon race that went awry in 1997, when infrared shock waves generated by the Concorde supersonic airliner led tens of thousands of birds, known for their navigational acumen, off course and many were never seen again.

A number of animals exclusively use the Earth's magnetic field as a compass. The list with a magnetic sense has grown to include species in every vertebrate category, as well as certain insects and crustaceans. Some may use it to orient, such as blind mole rats. Others including salmon, spiny lobsters, and nightingales may use it for navigation and homing, alongside other sensory clues.

The Earth has a magnetic field similar to that of a bar magnet, it is composed with an extremely hot solid inner core, two thirds the size of the moon made up primarily of iron. At 5,700 degrees Centigrade this iron is as hot as the Sun's surface, but the high temperature caused by gravity prevents it from becoming liquid. Surrounding this is the outer core, a 2,000 kilometer thick layer of iron, nickel, and small quantities of other metals. Lower pressure than the inner core means the metal here is liquid. Differences in temperature, pressure and composition within the outer core cause convection currents in the molten metal as cool, dense matter sinks while warm, less dense matter rises. This flow of liquid iron generates electric currents, which in turn produce a magnetic field with north and south poles that engulfs the planet.

One other new book, *Wayfinding*, written by M.R. O'Conner, is worth reading. *Wayfinding* in the simplest terms is the use and organization of sensory information from the environment to guide us. She writes about human navigation rather than animals and that collective storytelling and close observation were key to helping us find our way. Her study focuses on part of our brain, the hippocampus, a structure deep in the medial temporal lobe that is believed to be essential for maintaining one's bearing in space and time. O'Connor is concerned that our reliance on GPS turn-by-turn direction will negatively affect hippocampal function in the future. Both O'Connor and Barrie have written books designed to make readers think about what we lose when we blindly outsource navigation to GPS.

Termites

—*Surprising facts!*

ACCORDING TO LISA Margonelli in her latest book, *Underbug*, every story about termites mentions how they consume somewhere between $1.5 and $20 billion in U.S. property every year. Termites' offense is often described as the eating of "private" property, which makes them sound like anti-capitalist anarchists. While termites are truly subversive, it's fair to point out that they'll eat anything pulpy. They even find money itself to be tasty. In 2011 they broke into an Indian bank and ate $220,000 in banknotes. In 2013 they ate $65,000 that a woman in Guangdong had wrapped in plastic and hidden in a wooden drawer. Another statistic seems relevant: termites outweigh us ten to one. For every 132 pound human there are 1320 pounds of them.

Termites are indeed remarkable and are among the rare organisms that can feed on wood. This unique ability has allowed them to become one of the most abundant creatures in the tropical forests and they have the bacteria in their gut to thank. Many animals rely on the organisms in their guts to aid the process of digestion, as an example, there are approximately 39 trillion bacteria in the human gastrointestinal tract. However, the termite gut microbiome is among the most complex of any animal group. It comprises a diverse mixture of bacteria, protists (one celled organisms that are not plants, animals or fungi), protozoa and fungi which can break down and extract essential nutrients from materials that are indigestible to most animals. The analysis of a termite entombed for 100 million years in an ancient piece of amber has revealed the oldest example of mutualism ever discovered between an animal (a termite) and microorganism (a protozoa), and also shows the unusual biology that helped make this one of the most successful insect group in the world.

Termites are unusually surprising in another respect. In groups of a million or two, they are formidable architects, building mounds that can reach as tall as 20 feet or higher and 40 feet in diameter. Considering that termites are 0.4 inches in length, the human equivalent in terms of buildings would be structures up to 4.600 feet tall. Moreover, termites can burrow as far as 225 feet underground. The 33 pounds or so of termites in a typical mound, will, in an average year, move a fourth of a metric ton (about 550 pounds) of soil and several tons of water. Mounds are oriented north-south as accurately as if plotted with a compass, in order to maximize heat from the sun. Some termites protect the mound by spraying chemicals from nozzles on their heads at intruders.

Although known as "white ants", termites are not ants and their relationships with other insects remain unclear. Molecular analyses show that termites are social cockroaches, no longer meriting being classified as a separate order (Isoptera) from the cockroaches (Blattodea).

Termites thus join the more than 3,500 species of roaches worldwide, 55 of which are found in the United States. It is surprising to find that a group of wood feeding cockroaches has evolved full sociality. Termites are amazing in a number of aspects. Their queens are 30 times the size of normal soldiers and workers and produce about 30 eggs per minute to keep the colony alive. Queens lay millions of eggs in the course of their decades long life—the longest life span of any insect.

Termites are mainly known for damage caused to human beings, both in urban and rural areas. However, these insects play an important role in decomposing organic matter in tropical areas and are important natural resources, which are widely used in traditional medicine and are also consumed by human populations in several parts of the world. In Africa, Asia, Latin America and parts of Indonesia, eating termites or using them to feed livestock is a way of life. Collected at the start of the rainy season when other sources of protein are scarce, termites are best eaten after being slightly roasted. (If you are a gourmet, salted dried termites are available from Amazon.) Termites are among the most commonly consumed insects on the planet, second only to grasshoppers. The importance of insects as a food source for humans is not surprising, since this is the group with the highest number of species in nature. Entomophagy, as the practice of using insects as part of a human diet is called, has played an important role in the history of human nutrition. They have been consumed for generations , and this practice has increased in popularity in recent years. According to a publication in the Journal of Ethnobiology and Ethnomedicine, termites are also used in treatment of various diseases that affect humans, such as influenza, asthma, bronchitis, whooping cough, sinusitis, tonsilitis and hoarseness. Termites are indeed unique and surprisingly versatile.

THE BRAIN

—Much still to be learned and unlearned.

ACCORDING TO ORNSTEIN and Thompson's book, *the Amazing Brain*, "analogies related to the brain, such as comparing it to computers, switchboards, or any machine yet to be invented are grossly inadequate, for the brain is unique, and unlike anything man has ever made." Truer words were never spoken. Although scientists have made progress in learning how the elements and chemicals of the brain function—a great deal remains to be discovered. There is a lot more to it than its constituents, 75 to 80 percent water, with the rest split between fat and protein. These three substances come together in a way that allows thought and memory and vision and aesthetic appreciation of all the rest. Because of its makeup, the brain is surprisingly soft, not unlike soft butter. Brain surgery, therefore is an extremely delicate and demanding operation and the ability to perform it has long been considered the ultimate measure of human intelligence. Some surgeons in the past, however, have done their best to disprove that statement. One of them, Walter Jackson Freeman, actually became quite famous. He was influenced by work done by a professor of neurology at the University of Lisbon, Egas Moniz in the mid 1930s and the recipient of the Nobel Prize in 1949. Dr. Moniz experimentally cut the frontal lobes of schizophrenic patients to what he thought would quiet their troubled minds. (The frontal lobe part of the cerebrum, is primarily involved in motor function of voluntary muscles, learning, planning, decision making, and purposeful behavior.) He called the operation a leucotomy. Moniz, however, was not an advocate of the scientific method. There were no animal or cadaver studies, no hypothesis, no risk analysis or patient consent and little if any post-surgery follow up. His major impetus was the results he witnessed following surgery on one chimpanzee. Despite these shortcomings, Freeman modified the procedure renaming it a lobotomy (a lobotomy is defined as a surgery involving incision into the prefrontal lobe of the brain). With the help of a neurosurgeon James Watt, he performed the first prefrontal lobotomy in the United States on a 63 year old woman who was suffering from insomnia and agitated depression. The operation consisted of drilling six holes into the top of the woman's skull and when it was finished she emerged transformed and lived for another five years. He soon developed a more "efficient" way to perform the procedure by drilling into a patient's head. It involved rendering him or her unconscious by electroshock before inserting a standard household ice pick into the brain through the eye socket. The instrument would then be hammered into the skull and wiggled back and forth to sever the connections to the prefrontal cortex in the frontal lobes of the brain. His procedure often resulted in patients left in a vegetative state and he may have been responsible for an estimated 490 deaths.

One of the vegetative patients was Rosemary Kennedy, sister of President John Kennedy. She had learning disabilities and was subject to mood swings. In 1941, when she was twenty three, her father had her lobotomized by Dr. Freeman. It virtually destroyed her. Rosemary eventually regained the ability to walk but several times she wandered off on her own. She spent the next sixty four years of her life in a care home, unable to speak, incontinent, and without any semblance of personality. Her mother did not see her daughter for twenty years, and no family member visited her until 1958, when John Kennedy secretly paid a call while campaigning in Wisconsin.

As described in Jack El-Hai's book, The Lobotomist, Freeman's most controversial lobotomy patients were the children he treated—nearly all of them diagnosed with schizophrenia, and the controversy dogged him throughout his career. The book provides comprehensive details about Freeman and his ideas. It should be noted that despite much criticism about the procedure, it gained popularity through major publications across the country, hailing the lobotomy as a "miracle" surgery. By 1949, 5000 lobotomies were being performed annually, up from just 150 in 1945. Freeman would ultimately lobotomize more than 2900 patients, including 19 children younger than age 18. Long term studies on the effects, however, began to surface and many supporters began to abandon it. Freeman performed the last surgery in 1967 after severing a patient's blood vessel during the procedure, resulting in death three days later. It is sad, but Freeman clung to his belief for more than 35 years, from the age of shock therapy to the era of psychiatric drugs. El-Hai noted that while today's remedies are different, some still echo Freeman's work.

The Doomsday Clock

According to Wikipedia, the Doomsday Clock is a symbol which represents the likelihood of a man-made global catastrophe. It was conceived by University of Chicago scientists who had helped develop the first atomic weapons. They began the Bulletin of the Atomic Scientists in 1945 and two years later created the Doomsday Clock which has become the best known measure of humanity's risk of mankind's total annihilation, the apocalypse (midnight). Should an meteor large enough to destroy the earth be destined to land here, it would be, according to the bulletin's science and security board, doomsday.

Until 2007, it only tracked the danger of nuclear weapons, since then it has also incorporated the effects of climate change. Today the clock is set by nineteen members of the Bulletin's Science and Security Board, including national security experts, physicists, climate scientists, a public health expert and a cybersecurity expert; and the Board of Sponsors, including fifteen Nobel Prize laureates. Each year the bulletin decides whether to move the clock or leave it as is. The clock was set at 7 minutes until midnight in 1947. In 1949, when Russia tested its first atomic bomb the clock was moved 4 minutes closer, or 3 minutes until midnight. The high mark was 17 to midnight, set in 1991 after the US and Russia signed the Strategic Arms Reduction Treaty. Just recently it reached the lowest mark to date, 2 to midnight. The reasons are rising tensions between the US and Russia once again, and similar problems with China and North Korea. Other related issues include the continuing arms race between India and Pakistan, Iran's quest for nuclear arms and the persistent threat of climate change. The clock had been 2 to midnight once before, in 1953, when the US and Soviets tested thermonuclear weapons, within six months of each other. That year, Eugene Rabinowitch, a former Manhattan Project scientist who cofounded the Bulletin, wrote "The achievement of a thermonuclear explosion by the Soviet Union, following on the heels of the development of similar devices in America, means that the time, dreaded by scientists since 1945, when each major nation will hold the power of destroying, at will, the urban civilization of any other nation, is close at hand.

According to the bulletin, 2018 was the year where the failure of world leaders to address the largest threats to humanity is lamentable—but that failure can be reversed. This year marked two minutes to midnight, but the Doomsday Clock has ticked away from midnight in the past, and during the next year, the world can again move it further from apocalypse. The warning the Science and Security Board now sends is clear, the danger obvious and imminent. The opportunity to reduce the danger is equally clear. The world has seen the threat posed by the misuse of information technology and witnessed the vulnerability of democracies to disinformation. But there is a flip side to the abuse of social media. Leaders react when citizens insist they do so, and citizens around the world can use the power of the internet to improve the long-term prospects of their children and grandchildren. They can insist on facts, and discount nonsense. They can demand action to reduce the existential threat of nuclear war and unchecked climate change. They can seize the opportunity to make a safer and saner world.

The year before, the board warned that the probability of global catastrophe is very high, and the actions needed to reduce the risks of disaster must be taken very soon. It is two and a half minutes until midnight, the clock is ticking and global danger looms. Wise public officials should act immediately, guiding humanity away from the brink. If they do not, wise citizens must step forward and lead the way.

In 2015, the board announced that unchecked climate change, global nuclear weapons modernizations, and outsized nuclear weapons arsenals pose extraordinary and undeniable threats to the continued existence of humanity, and world leaders have failed to act with the speed or on the scale required to protect citizens from potential catastrophe. These failures of political leadership endanger every person on earth. (Three minutes to midnight.)

Chronologically, it does not appear that much progress has been made in the past five or six years. In 2012, it was five minutes to midnight.

The clock, however, does not lack for critics or criticism, some say that warning people of danger causes a political paralysis. Others question the judgment of the expert panel. Overall, however, the clock is a potent symbol of the scientific concern for humanity's possible annihilation and allows political leaders to make decisions actually based on facts and to manage the dangers—not a bad idea.

The Epizootic of 1872

—A surprising series of events!

THOSE OF US interested in science and medicine are acquainted with the Spanish flu of 1918, the disease that killed as many as 100 million people, more than both world wars combined. The word "influenza" began its use toward the end of the Middle Ages, stemming from the Italian for "influence"—the influence of the stars. The Spanish flu and the Black Death of 1348 are two exceptionally deadly diseases that threatened the very existence of mankind. The Black Death may have decimated as many as 60 percent of the population living in the world at that time, an estimated 50 to 200 million people. Much has been written about both pandemics and their devastating effects on society and the economy. Strangely little has been published on a history making disease that plagued the United States and Canada in 1872. It, however, did not affect humans, but horses, and was later termed an equine influenza. (Widespread diseases in animals are called epizootic, while epidemic or pandemic are the words used for widespread disease in people.) The epizootic disease was characterized in this rather simplistic poem:

> Not a sound was heard in the silent street,
> As home from the concert we hurried;
> For we found not a street car, carriage or bus,
> And we felt considerably worried.
>
> We hailed a driver we used to know
> And hurriedly asked the reason;
> He said, as he sadly shook his head,
> That the horses were all sneezin.

In 1872, horses were like cars and trucks of today. Horses were necessary to move people and products; without horses, commerce would grind to a halt. In late September, a mysterious illness swept through the horse population in Toronto, Canada. It possibly originated in farms outside of Toronto and culminated with fourteen sick horses in one city stable. Then according to one reporter, the disease "ran through the city with remarkable rapidity, sparing scarcely a single animal of this noble species." There were sick horses everywhere, suffering from hacking coughs, respiratory distress and fatigue. Fortunately few horses died. Surviving horses were unable to work for two weeks or more. This outbreak was the beginning of what came to be known as the Great Epizootic. Following the events in Toronto, the disease spread throughout North America from the Atlantic to Pacific coasts and into parts of Central America. It was the most explosive animal disease ever documented and temporarily suspended transport and trade. Moreover, it was responsible for the infamous fire that consumed Boston in November 1872 because fire equipment was useless without the labor of horses. And it slowed voters on the eve of the election of Ulysses S. Grant. Some cities replaced horses with other animals, including oxen, mules and even people.

Every city the epizootic visited experienced the symptoms of influenza that incapacitated horses, suspending street railway service and local deliveries, temporary shortages of food and other supplies, price gouging, and an inability to arrest the spread of the disease. The disease found ideal conditions in the 1870s: cities filled with thousands of horses kept in cramped, crowded stables. The conditions were environmentally similar when horses were subjected to travel by railroad. As mentioned earlier, horses were the predominant mode of interurban transportation in 19th century cities and lived by the thousands clustered in stables from Montreal to New York City to Galveston, Texas. No large city in 1872-3 relied on any other domestic animals for transportation and other labor more than the horse. Nineteenth century cities in both Canada and the United States shared some common ecological characteristics and structures including the role of the horse. Cities constituted habitats for large numbers of humans and livestock animals. These common characteristics left North American cities vulnerable to the outbreak of animal diseases.

According to a recent article in Environmental History, the Great Epizootic moved like wildfire, burning most fiercely wherever it found ample fuel in the form of horse bodies and the means of reaching those bodies, usually by railroad. The account of the movement, based on analysis of over 480 newspaper accounts and reports published between 1872 and 1873, indicated that the disease appeared in 164 Canadian and US cities and towns, affecting nearly all horses. There are very detailed maps that show the course of travel of the illness included in the article.

It is of interest that in the 19th century, many veterinarians attributed diseases to bad air or filth poisoned by the decomposition of animal and plant matter. However, examining how the disease moved from city to city led experts to conclude that its path was forged by some form of contagion and that logical proof indicates that epizootic influenza spreads only by virtue of its communicability. This was another major step forward toward understanding disease transmission.

♥ The Heart

—A symbol and its function.

The symmetrical heart symbol is a far cry from the actual fist sized mass of tissue we all carry inside of us. Its earliest illustration was created around 1250 in France and somehow became associated with romance. The shape evolved during the 14th century, starting with images in the work of an Italian artist. Anyone in love knows that your heart beats faster when you see someone who sparks your romantic interest. We see examples of the symbol everywhere from jewelry to emojis to Valentine's day and beyond. The heart icon is everywhere. According to a recent article in the New York Times, the symbol's appeal is because its shape, with two halves forming a single figure, so perfectly captures the Platonic idea of love as the longing to merge with an ideal soul mate.

In truth, the heart looks nothing like its symbol, it is a lumpy muscular organ that functions as a circulatory pump for our body. It weighs around eleven ounces and takes in deoxygenated blood through the veins and delivers it to the lungs where it gets rid of carbon dioxide and picks up fresh oxygen before pumping it into the various arteries. During an average lifetime, a heart pumps enough blood to fill 13 supertankers, each ship capable of holding one million barrels, each barrel containing 42 gallons of blood. The heart sits between the lungs with its top tilted toward the right side of the body. Most of the heart consists of muscle, known in medical terms as the myocardium. The interior of the heart harbors two pairs of hollow chambers, each pair contains a small antechamber—an atrium, or auricle—and a larger section called a ventricle. As might be expected, understanding heart function is not simple, considering its criticality. The life of each of our seventy five trillion cells depends on its uninterrupted beats. It was not until the third decade of the 17th century, that William Harvey discovered that the heart really works as a pump and that veins contain valves that permit flow of the blood in a single direction.

How important is the heart? According to the Center for Disease Control and Prevention about 610,000 people die of heart disease in the United States every year—that is 1 in every 4 deaths. It is the leading cause of death for both men and women. More than half of the deaths due to heart disease in 2009 were in men and coronary heart disease kills 370,000 people annually, it is the most common type of heart disease. Moreover, every year about 735,000 Americans have a heart attack. Of these, 525,000 are a first attack and 210,000 happen in people who have already had a heart attack. Heart disease is also the leading cause of death for people of most ethnicities in the United States, including African-Americans, Hispanics and whites. For American Indians or Alaska natives and Asians or Pacific Islanders, heart disease is second only to cancer. High blood pressure, high cholesterol and smoking are key risk factors for heart disease.

About half of Americans have at least one of these three risk factors. Several other medical conditions and lifestyle choices can also place people at a higher risk for heart disease, including: diabetes, overweight and obesity, poor diet, physical inactivity and excessive use of alcohol.

We continue to learn more about other risk factors as well. While doctors routinely test for certain types of cholesterol, like high density lipoprotein (the good), low density lipoprotein (the bad), few test for lipoprotein (a), also known as Ip(a), high levels of which triple the risk of having a heart attack at an early age. Lipoprotein was discovered in 1963, but no one knew precisely what purpose it serves in the body. It was thought to have beneficial role in repairing injured cells, but it has been found to accelerate the formation of plaque in the arteries which promotes blood clots. In a recent New York Times editorial, the author reported that few doctors know about Ip(a). The author suggested that testing for Ip(a) should be considered for people with an early onset of cardiovascular disease—which means younger than age 50 for men and age 60 for women—or a strong family history of it. Once Ip(a) is identified, physicians should mitigate its effects by aggressively lowering LDL cholesterol, lowering blood pressure, reducing blood sugar, and encouraging healthy diets and exercise. Sound advice for all of us.

The Periodic Table

—Something to know more about.

ANYONE WHO STUDIED chemistry likely reluctantly remember the Periodic Table. It is one of science's greatest creations and became 150 years old this month. How it was developed and its significance are perfect illustrations of the process of scientific progress. Even if you know little about chemistry, the periodic table has had an effect on you by playing a major role in your life and lifestyle.

What is the Periodic Table? There are over one hundred elements and they are usually presented to reflect their detailed structure. Elements are the simplest chemicals. Examples include hydrogen, oxygen, iron and nitrogen. (Hydrogen and oxygen combine to form water.) An element cannot be made any simpler because it only contains one type of atom. There are over one hundred different types of atoms, and so elements. Most of them are metals. The rest are called non-metals.

The periodic table is organized in the order of the number of protons in the nucleus, which is called 'the atomic number'. This means that hydrogen is the first element with one proton in each nucleus, helium two, lithium three and so on. The Periodic Table is nature's Rosetta stone. To the uninitiated, it's just 100-plus numbered boxes, each containing one or two letters, arranged with an odd, skewed symmetry. To chemists, however, the periodic table reveals the organizing principles of matter, which is to say, the organizing principles of chemistry. Some like gold, iron and sulfur, had been known since ancient days. Others, like manganese, molybdenum, and tungsten were recent discoveries. John Dalton, an Englishman, showed that elements reacted in fixed proportions by weight, but also that those proportions were ratios of small whole numbers. The simplest way to explain this was to suppose each element to be composed of tiny, invisible particles, all of the same weight. The Greek word for indivisible is "atomos". Thus the atom was born. Dalton's work was followed by contributions by Jacob Berzelius, a Swede. He furnished chemistry with its language while coming up with the idea of abbreviations that now occupy the periodic table's rectangles. Back in England Sir Humphrey Davy employed a newly developed battery to discover sodium and potassium in 1807 and magnesium, calcium, strontium, barium and boron in 1808. He also showed that chlorine, previously thought to be a compound of oxygen, was actually and element. After Davy's findings, new elements were being found more frequently—- iodine (1811), cadmium and selenium (1817), lithium (1821), silicon (1823), aluminum and bromine (1825). By then there were enough elements for the next steps in the process of creating the table.

As more and more elements turned up, the search for some order intensified. In 1864, John Newlands, an

Englishman, published what is known as the law of octaves. Arranging the known elements in order of atomic weight. Dimitri Mendeleev, however, is primarily credited with the table's inception. According to a recent article in The Economist, Mendeleev looked like a mad professor, his hair was cut just once by a shepherd using wool shears. He also behaved like a mad professor, prone to dancing rages. On the other hand, he was consummately patient and arranged the 63 then known elements in a pattern, arguably the most important game of patience ever played. Gaps in his table were subsequently added. Radon was added by William Ramsey, another Brit. With various collaborators he added argon in 1894, helium in 1895 and neon, krypton and xenon in 1898. Ramsey used the newly developed technology of cryogenics, which he used to liquefy air and then separate it into its components, according to their boiling points.

Fast forward to the 1930s, when physicists discovered that radioactivity could in essence be reversed by bombarding atoms with subatomic particles to increase their atomic numbers. This way, new elements were produced. Technetium, created in 1937, was the first. Many others including francium, actinium followed. From then the extension of the table became work of physicists, not chemists.

At a fundamental level, all of chemistry is contained in the periodic table and chemistry is an integral part of all of our lives. There are many examples of chemistry in our daily lives, for example, digestion relies on chemical reactions between food and acids and enzymes to break down molecules into nutrients the body can absorb and use. Cooking is a chemical change that alters food to make it more digestible. The heat of cooking denatures proteins, promotes chemical reactions between ingredients and carmelizes sugar. Drugs work because of chemistry, the chemical compounds attach to binding sites for natural chemicals in our body.

The Science of Cooking

—Make it more fun.

My wife has gratuitously given me the opportunity to prepare the evening meal much to her chagrin and I am having fun learning about the entire process by treating it like a chemistry experiment. As a guide I have consulted with Howard Hillman's book, *Kitchen Science* and *Culinary Reactions: The Everyday Chemistry of Cooking* by Simon Quellen Field. According to the latter, cooking is often about combining ingredients to create something completely different. It involves many chemical and physical changes to the food that the cook carefully controls in order to produce the desired result. Nowhere is this more true than baking. From as early as high school we may have learned that carbon dioxide (CO_2) bubbles are generated whenever water is poured over a dry acid and alkali mixture. This is exactly what occurs when we use baking powder because this cooking ingredient is a blend of acid (calcium acid phosphate, sodium aluminum sulfate or cream of tartar) and alkali (sodium bicarbonate, commonly known as baking soda). Add water or some other liquid and a chemical reaction results producing CO_2. This generated gas creates minuscule air pockets, or enters into existing ones, within the dough or batter. When placed in a hot oven or on a hot griddle, the dough or batter rises, for two reasons. First, the heat helps release additional carbon dioxide from the baking powder, and second, the heat expands the trapped carbon dioxide gas and air and creates steam. The resulting pressure swells the countless air pockets, which in turn expand the baked food making it lighter and more delicious.

Yeast (*Saccharomyces cerevisiae*) is also used to leaven bread (to make it rise). It is made up of minute one cell fungi that rapidly multiply if given their favorite foods of sugar or starch in a moist environment. When the yeast cells feast on the sugar or starch that a yeast enzyme has converted to glucose (sugar), a chemical reaction takes place. The sugar ferments, and as a result, is converted into alcohol and carbon dioxide, along with trapped air and steam which leavens the bread.

Chemistry is also responsible for the taste of tomatoes. This time of year when they are so plentiful, it is wise not to refrigerate them. Tomatoes won't be as aromatic and savory because cold hinders the conversion of the vegetable's linolenic acid to Z-3 hexenel, the compound that accounts for much of the desirable ripe tomato scent and taste. Cold also reduces the volatility of molecules and therefore the number of Z-3 hexenel molecules that reach the olfactory receptors in the nose.

The density of a mixture is an important factor when cooking. Density is measured by floating a little scale (called a hydrometer) in the water, which tells how much sugar, alcohol, and water are in the mix at any given time. A density test can also tell you how fresh your eggs are.

Place an egg in water, then dissolve measured amounts of salt into the water until the egg floats. A bad egg will float right away.

Another place where density comes into play in the kitchen is in making hard-boiled eggs. The yolk of an egg contains fats and oils and is thus less dense than the white of the egg. This means that if left to itself, the yolk inside will float to the top of the egg and thus be off-center when the egg is cut in half for deviled eggs or sliced into a salad. To keep the yolk centered, the eggs must be turned frequently while being cooked, keeping the yolk away from the shell. Since the white of the egg cooks on the outside first (where it is closer to the boiling water), the yolk that is turned often will not be able to get past the hardening white and will end up centered.

Other chemical reactions are involved in whipping creams and beating egg whites. In both of these processes proteins in the foam are first denatured, which, as the name implies, means that they are changed from their natural state. Proteins are made up of building blocks called amino acids. Some of these building blocks are attracted to water but avoid oils and fats. Others are attracted to oils and fats but are repelled by water.

In whipped egg whites, you get bubbles with a protein film. The water-loving parts stick into the water, and the water-avoiding parts stick into the air bubble. In whipped cream, you get big bubbles of air surrounded by a film of protein, surrounded by tiny globules of fat stuck to the fat-loving parts of the protein, connected to another film of protein that forms the wall of the next bubble.

FINAL THOUGHTS

In essence, cooking provides a greater appreciation for science and makes it less of a chore. Science does the same for cooking. Try it some time.

The Scientific Revolution

I HAVE BEEN watching a series of lectures sponsored by Hillsdale College, the most recent covered the scientific revolution and was quite fascinating. Science has been described in numerous ways, the best definition may be that is the study of the nature and behavior of natural things and the knowledge that we obtain about them. The knowledge is based on facts learned through experiments and observation. Physics, chemistry and biology are examples of a science. The word "scientist" was not coined until 1833, when William Whewell described that person as an expert in the study of nature. Whewell was a Cambridge professor, Master of Trinity College, theologian, philosopher and historian. Prior to then, scientists were called natural philosophers.

The scientific revolution may have begun with three publications somewhere around the mid 1500s. They were instrumental in moving civilization from the dark ages to what historians refer to as the Renaissance. One was the translation from ancient Greek to Latin of Archimedes work written around 200 BC. The second was Andreas Vesalius seminal description of the human body, and the last was the revolutionary astronomical discoveries by Nicolaus Copernicus. The revolution may be considered closed in 1704 when Isaac Newton published his *Opticks*.

Archimedes

Archimedes was a Greek mathematician, philosopher and inventor, who wrote important works on geometry, arithmetic, and mechanics. He began the science of hydrostatics, which deals with the pressure of liquids. Hydrostatics are sometimes known as "Archimedes principle" stating that a body immersed in fluid loses weight equal to the weight of the amount of fluid it displaces. He discovered the laws of the lever and pulleys, which led to machines that could move heavy loads, or increase speeds, or change directions. He discovered the principle of buoyancy, which tells us why some things float and some things sink and some things rise into the air. Archimedes discovered that every element, and even every combination of elements, has a different density, or weight for its size—and that this is a good way to tell one substance from another, even if they look alike. He invented the Archimedean screw, a device that is still used to drain or irrigate fields and load grain and run high-speed machines and a type of astronomical machine that showed eclipses of the sun and moon. He also estimated the length of the year, and the distances to the five planets that were known to the ancient world.

Vesalius

According to author Sherwin Nuland, Andreas Vesalius wrote the book that epitomizes the confluence of science, technology and culture in a way that few, perhaps no other books have every done. It was an outgrowth of vigorous spirit of the Renaissance and celebrated a return to the logical thoughts and observational methods of the ancient Greeks like Archimedes. *De Humani Corporis Fabrica* paved the way for modern scientific medicine by presenting the world with the first accurate description and knowledge of human anatomy and a method by which it can be studied. Even the language of the text, a brilliant form of Latin is reminiscent of the finest of Roman rhetoric. With the book's publication, medicine was finally lifted out of the medieval murkiness represented by much earlier so-called anatomists. The book was technically accurate, based on Vesalius' dissections of human cadavers and annotated with remarkable illustrations that brought anatomy to life. The illustrations were done by a pupil of Titian. Unfortunately the text of Fabrica remains the least read of the great books of medicine, and to this day only fragments of it have been translated into English. Nevertheless, modern medicine owes much to his dissections.

COPERNICUS

Nicolaus Copernicus' book *De Revolutionibus Orbium Coelestium*, published in 1543 gave the world the most important scientific insight into the modern age. (He was seventy years old at that time.) It was the radical theory that the earth and other planets revolve around sun (the heliocentric model). Prior to Copernicus nearly everyone believed that a perfectly still earth rested in the middle of the cosmos, where all the heavenly bodies revolved around it. Copernicus was also the first to proclaim that the earth rotates on its axis once every twenty four hours. He began his research around the year 1510, but concealed the theory for thirty years, fearful of ridicule from his mathematician peers. Decades after his death in 1543, when the first telescopic discoveries lent credence to his intuitions, the Catholic Church's Holy Office of the Inquisition condemned his efforts. In 1616, On the Revolutions was listed on the Index of Prohibited Books, where it remained for more than two hundred years. The philosophical conflict and change in perception that his ideas engendered are sometimes referred to as the Copernican Revolution.

Final thoughts

According to David Wootton in his book, The Invention of Science, " the vast scientific revolution has transformed the nature of knowledge and the capacities of humankind. Without it there would be no Industrial Revolution and none of the modern technologies on which we depend; human life would be drastically poorer and shorter and most of us would live lives of unremitting toil."

The two smartest men who have lived in the past 400 years.

Now that I am retired I spend my spare time in odd ways. For example, I have often wondered why there are so many people smarter than I am and what makes them so. This bizarre curiosity led to my admiration and envy of many individuals. However, to narrow down the list to something more reasonable, I elected to choose two people whom I deem the most intelligent and limit the time frame to the past 400 years. My criteria are their diversity of knowledge, versatility, perceptions of others, achievements, publications and legacies. The roster could include a number of stalwarts including Isaac Newton, Albert Einstein, Johann Goethe and many more. I chose two who are not as well known and in my opinion Leonhard Euler (pronounced Oiler) and Thomas Young to top the list. This may seem odd, and many people have not heard of either one. Let me explain. Both gentlemen were experts in multiple fields, and their works and concepts are an integral part of modern engineering, astronomy, medicine, physics, and other fields as well. Euler derived modern mathematical principles and contributed to the fields of geometry, trigonometry and calculus. In calculus alone, he provided hundreds of discoveries and proofs along with many computations to simplify (?) and clarify differential calculus, infinite series and integral techniques. He was a revolutionary thinker in such diverse fields as astronomy, acoustics, hydrodynamics, mechanics, music, ballistics, navigation and topology. As one would expect, Euler had a remarkable memory, at an early age he could recite all 9800 lines of Virgil's poem, the Aeneid. His productivity was equally amazing, during his career he wrote more than 850 publications, including 18 books. Euler lost sight in one eye in his early 30s, and was nearly blind by age 60. Despite that, he continued his illustrious career and published more than 400 more articles and a major three volume work on lunar motion. Pierre Simon LaPlace, a noted 18th century mathematician, reportedly said, "Read Euler, read Euler, he is the master of us all."

Thomas Young was another polymath, a linguist, physician and physicist who established the theory of light, color perception, anatomy, the significance of energy, elasticity and because of his uncanny knowledge of language, the study of Egyptology and hieroglyphics. The latter was instrumental in his ability to help translate the previously undecipherable and mysterious Rosetta stone. I first became interested in Young because of my background in pharmacy and mechanical testing, finding it hard to believe that Young's Rule—a method to determine a child's dosage for drugs, and Young's Modulus—a measure of stiffness of an elastic material used by all material scientists are both derived by the same person. They were. Young also worked on liquid molecule size and surface tension measurement. While he was still a medical student he discovered how the lens of the eye changes shape to focus on objects at different distances which led to the discovery of the cause of astigmatism. When asked in later years to contribute to a new edition of the *Encyclopedia Brittanica,* Young offered to write on: the alphabet, annuities, capillary action, cohesion, color, dew, Egypt, focus, friction, haloes, hieroglyphics, hydraulics, motion, resistance, ships, sound, strength, tides, waves and anything about medicine. He requested ,however, that this contributions be kept anonymous. His wish tells you something about his character.

Many will not agree with my choices and justifiably so. An author of a recent book I read chose two others, one old and one fairly recent. In his mind Leonardo Da Vinci and physicist Richard Feynman were the smartest, based on their curiosity and achievements. Leonardo's boundless interests spanned such broad swaths of art, science and technology and he remains to this day the quintessential Renaissance man. Leonardo, however, was born in 1452 and doesn't meet my 400 year requirement. Feynman's genius and achievements in numerous branches of physics are legendary, but he also pursued fascinations with biology, painting, safe cracking, and studying Mayan hieroglyphs. He also wrote a number of best selling books.

In summary, I realize how presumptuous I am choosing two individuals as the most brilliant considering the billions of people who have inhabited or now inhabit the earth. There is no analytical measurement, and there were no intelligence tests available in the 18th century. However even if there were IQ tests, they would not be the only measure of genius. There seems to be no common denominator except uncommonness. My selections are based on and determined by the knowledge Euler and Young accumulated during their lifetimes and, by good fortune, their concerted efforts to share it with the world. Who would you pick?

Max Sherman, Warsaw, Indiana.

Ticks

—*Much more dangerous than we thought.*

Residents in the Midwest are generally aware that ticks carry the bacteria that cause Lyme disease. But ticks are responsible for other fatal diseases as well. The worst is Powassan disease, which kills about 10 percent of its victims and leaves half of the survivors with permanent neurological damage.

Ticks carry other pathogens that cause such human diseases as anaplasmosis, babesiosis, ehrlichiosis, Rickettsia parkeri, Rocky Mounted Spotted Fiver, Southern Tick Associated Rash Disease, tick-borne relapsing fever, tularemia and Rickettsiosis. Like mosquitoes, ticks are capable are capable of carrying, supporting and injecting more kinds of disease-causing microbes than almost any other creature..

Description

Ticks are close relatives of mites and spiders. In fact, they may have evolved from mites more than 94 million years ago.

Ticks have eight legs and a flat, hard body. Like many of their arthropod relatives, ticks hatch from eggs and grow through four distinct stages: egg, six-legged larva, eight-legged nymph and adult. Adult ticks are approximately the size of a poppy seed. Ticks, like crabs and lobsters, have a hard covering that they must shed periodically to grow. They generally live for two years.

The biggest problem ticks have is finding an appropriate host for their next blood meal. (Ticks must have a blood meal at every stage to survive and grow.) Ticks cannot fly; they crawl slowly and generally cannot travel without help. Ticks can crawl on a blade of grass or twig and using their lower legs for leverage wait for the right host to come along. They hold their upper pair of legs outstretched, waiting to climb aboard.

Ticks do not see well but have an extremely sensitive and rapid response to a whiff of carbon dioxide or to the faint odor of butyric acid exuded from the skin of many animals. They can feed from mammals, birds, reptiles and amphibians.

Depending on the tick species and its stage of life, preparing to feed can take from 10 minutes to two hours. When the tick finds a feeding spot, it grasps the skin and cuts into the surface. The tick then inserts its feeding tube. Many species also secrete a cement-like substance that keeps them firmly attached during the meal. The feeding tube can have barbs that help keep the tick in place. Ticks can also secrete small amounts of saliva with anesthetic properties so that the animal or person cannot feel that the tick has attached itself.

Some ticks will attach to a host and suck the blood slowly for several days. If the host has certain blood-borne infections, such as the *Borrelia burgdorferi* spirochete, the agent that causes Lyme disease, the tick may ingest the pathogen and become infected. Once infected, a tick can transmit infection throughout its life. Transmission of *B. burgdorferi* generally requires at least 36 hours of tick attachment.

Lyme disease is the most widespread tick-borne disease in the United States, but it is not confined to this continent. Researchers have even located infected ticks that helped spread disease on seagulls and albatrosses in the Arctic Ocean and in Antarctica. Some 30,000 cases are reported to the Centers for Disease Control (CDC) each year, but most cases go unreported because the symptoms are mild or mimic other diseases. The CDC recently estimated that there may be 300,000 cases a year in this country.

Many scientists contributed to today's understanding of Lyme disease but a team from the Yale School of Medicine is widely credited with the discovery. Dr. Stephen Malawista and Allen Steere, then a postdoctoral student, defined the ailment.

The story began in 1975, when two mothers—one from Lyme, Connecticut, and the other from adjacent Old Lyme, were distressed by the odd rashes, neurological symptoms and swollen joints that their families and others were experiencing. Unable to find answers, each approached Yale independently. The initial diagnosis was juvenile rheumatoid arthritis, but the disease had never been known to appear in bunches. Doctors at Yale counted 51 cases, a rate about 100 times what was expected to occur in a combined population of 12,000 in the two towns. The cases also occurred almost exclusively in warm weather months. Because the disease was clustering, it looked like there had to be a vector—something like an insect to transmit the disease.

Malawista and his associates made the compelling link between ticks and the disease by noting that cases were 30 times more frequent on the east side of the Connecticut river, where Lyme is situated, than on the west side. Ticks feed and breed on deer, and there are far more deer on the east side.

Several points in the transmission cycle provide opportunities to prevent Lyme disease. However, no prevention strategy can be effective unless people who are at risk accept it. People can be cautioned to avoid tick-infested areas. Second, there is good evidence that removal of ticks within 36 hours after attachment will reduce the risk of infection. Daily tick checks are thus advised. People can also wear protective clothing, tuck their pants into their socks and use repellents.

For tick control, the most consistently effective method is to spray or otherwise broadcast acaricides onto vegetation where the ticks live. Acaricides are pesticides that kill ticks and mites.

Touch

—*A new science?*

I HAVE ALWAYS been intrigued with the five senses and as I grow older have suggested that "dignity" be added, if only because this quality appears to be on the decline. My earlier articles have described smell, sound and vision , leaving only taste and touch. This time I chose the latter because of something I read in the New Yorker earlier this year. Touch, according to the author, has become a new science and neuroscientists continue to learn more about how skin is busily sensing our relationships to others. The study is rapidly gaining new advocates who often refer to it by the term "haptics" which is Greek for touch. Touch is highly complex and closely related to the other senses as well as to the emotions. It is the unsung sense, the one we depend on most and talk about least. All of the senses can be, and have been thought of as having tactile dimensions.

Touch lies at the heart of our experience of ourselves and the world yet it often remains unspoken. The omission of tactile experience is noticeable not only in the field of history, but across the humanities and social sciences. We have so often been warned "not to touch" that we are reluctant to probe the tactile world even with our minds.

The historical study of touch has also been slighted because of the difficulty in coming to terms with the sense. A study is made difficult because it is at the same time the most complex and the most undifferentiated of the senses. Sight, hearing, smell and taste all have specific, limited sensory organs, all of which have specific limited functions. By contrast, skin is not only an organ of sense but it serves as the canvas upon which we "see" touch and its cultural associations.

Skin, our largest and most visible organ, is the flexible, continuous covering of the body that safeguards our internal organs from the external environment. This covering protects us from the attack by physical, chemical, and microbes and shields us from the most harmful rays of the sun while working hard to regulate our body temperature. The skin is constantly at work as a watchful sentinel, letting some things in and others out and home to hundreds of millions of microorganisms which feed on its scales and secretions. Skin is the interface through which we touch one another and sense much of our environment. Our skin reflects our age, our ancestry, our state of mind, our cultural identity, and much of what we want the world to know about us. Skin color is one of the ways in which evolution has fine tuned our bodies to the environment acting as a gradient to the intensity of the ultraviolet radiation that falls on different latitudes of the earth's surface. No other organ in the body can boast of so many diverse and important roles.

The sense of touch is controlled by a huge network of nerve endings and touch receptors in the skin known as the somatosensory system. It is responsible for all of the sensations, including cold, hot, smooth, rough, pressure, tickle, itch, pain, and vibrations. Within the system there are four types of receptors: mechanoreceptors, thermoreceptors, pain receptors and proprioreceptors.

Mechanoreceptors perceive sensations such as pressure, vibrations and texture, the most sensitive ones are found in the top layers of the dermis and epidermis and are generally found in non-hairy skin such as the palms, lips, tongue, soles of the feet, fingertips, eyelids, and the face.

Thermoreceptors perceive sensations related to the temperature of objects the skin feels. They are found in the dermis layer of the skin and include two categories—hot and cold. Thermoreceptors are found all over the body, but cold receptors are found in greater density than heat receptors. The highest concentration of thermoreceptors can be found in the face and ears (which explains why the nose and ears get colder faster than the rest of the body on a chilly winter day).

Pain receptors detect pain or stimuli that cause damage to the skin and other body tissues. There are more than 3 million throughout the body, found in skin, muscles, bones, blood vessels and some organs. They can detect pain caused by mechanical stimuli, burns, or chemicals (i.e., poisons from an insect bite). Pain receptors help to keep the body safe from serious injuries or damage by sending early signals to the brain.

Propiorecptors sense the position of the different parts of the body in relation to each other and the environment. They are found in tendons, muscles and joint capsules. This location in the body allows these special cells to detect changes in muscle length and tension. Without propioreceptors, we could not feed or even clothe ourselves.

Touch has widespread distribution, while the sensory receptors for sight, smell, taste and hearing are clustered together in the head close to the brain, touch receptors are scattered throughout the skin and muscle tissue and convey their signals via the spinal cord.

Tuberculosis

—A continuing global problem

In a Remember When column written a few months ago, Kristina Smiley described an event that occurred 50 years ago pertaining to required tuberculosis (TB) testing in Warsaw Community Schools. At that time, following the discovery of a single case, letters were sent to the parents of all students who were considered to have come into contact with that student. Tine tests were administered to 178 students and were negative except for one individual. Such testing is not routinely done today and the disease does not receive much attention in the United States. It is considered an ancient disease by most of us. However it is the leading cause of infectious disease and continues to be a major challenge to global health. According to the National Institutes of Health, each day, roughly 5000 people die of TB disease, resulting in nearly 2 million deaths in 2016 alone. Surprisingly, more than 1 billion people died of TB during the last 200 years, more deaths than from malaria, influenza, smallpox, HIV/AIDS, cholera and plague combined.

Tuberculosis is caused by a bacterium called *Mycobacterium tuberculosis*. The bacteria usually attack the lungs, but TB bacteria can attack any part of the body such as the kidney, spine and brain. Not everyone infected with TB bacteria becomes sick. As a result, two TB-related conditions exist: latent TB infection (LTBI) and TB disease. If not treated properly, TB disease can be fatal.

Tuberculosis bacilli are among the oldest and most widespread on Earth; and the liability to tubercular infection long antedated the emergence of humanity itself. Stone Age and Egyptian Old Kingdom skeletons have been diagnosed as exhibiting signs of tubercular damage. It has been hypothesized that the genus Mycobacterium originated more than 150 million years ago. In the middle ages, scrofula, a disease of the cervical lymph nodes, was described as a new clinical form of TB. The illness was known in England and France as "king's evil", and was widely believed that persons affected could heal after a royal touch. In 1720, for the first time, the infectious origin of TB was conjectured by the English physician Benjamin Marten, while the first remedy against TB was the introduction of the sanitarium cure. The famous scientist Robert Koch was able to isolate the tubercle bacillus and presented his result to the society of Physiology in Berlin on March 24, 1882. In the decades following this discovery, tuberculin skin tests, two partially effective vaccines, and anti-tuberculous drugs were developed.

'Tuberculosis' is an easy shorthand for pulmonary tuberculosis, from which most of the tuberculous suffer, although the disease is horribly ubiquitous. In addition to the lungs, it destroys the tissues of most of the body's systems—central nervous, circulatory, lymphatic, gastrointestinal, genitourinary as well as the bones, joints, and skin. Something of an oddity among the infectious diseases, tuberculosis has an undefined incubation period (from contact to symptoms), and a typically chronic course. TB once killed more people than any other single disease; the danger of death from consumption, as pulmonary tuberculosis was commonly called, tracked the footsteps of young men and women and loomed over courting couples for as long as they lived. Contracted in childhood or adolescence, tuberculosis left its victims more vulnerable to chronic or wasting illness in later years. In the middle of the nineteenth century diseases of the lung caused up to 25 percent of all deaths reported, and early in the twentieth century, when Americans were generally healthier, tuberculosis continued to head the list of fatal contagious diseases although though it resulted in fewer deaths. Even when tuberculosis had slipped to seventh place in the roster of fatal diseases by 1930, it remained terribly destructive—-the most frequent cause of death or disability during the critical ages of fifteen through forty-five. During that time, the uncertainties of treatment also magnified the public's fear of contagion. When a man was found to have tuberculosis he received a notice to quit his job and schoolchildren with the disease were ordered to stay home indefinitely.

Unfortunately, due to its infectious nature, chronic progression and the need for long-term care, TB continues to be a worldwide health problem. In more recent years, the appearance of multi-drug resistance and the current human immunodeficiency virus (HIV) epidemic, have presented additional challenges in designing therapy. (HIV weakens the immune system increasing the risk of TB in patients with HIV.) Multi-drug resistance has been a global problem and it now threatens the inhabitants of countries in Europe, Asia, Africa, and the Americas. There is currently an effort being made by the World Health Organization and the US Government to address fundamental TB research and to stimulate applied efforts to develop new diagnostic and therapeutic treatment methods.

Vanishing birds

—Something to be alarmed about!

Now the summer came to pass
And flowers through the grass
Joyously sang,
While the tribes of birds sang.

Walther von der Vogelweide's poem, *Dream Song*, written almost 900 years ago extols the virtues of the wide variety of songbirds—birds that are one of earth's treasures. It would be difficult imagining a world without them. Rachel Carson in her prophetic book, *Silent Spring* in 1962, wrote about losing birds thusly: "On the mornings that had once throbbed with the chorus of robins, catbirds, jays, wrens, and scores of other bird voices, there was now no sound." Her concern was the widespread use of pesticides and scientists this past month have documented that North America since the 1970's has lost nearly 3 billion birds, about 30 percent of the total. Pesticides are only partly to blame. Findings raise fears that some familiar species could go the way of the passenger pigeon, a species once so abundant that its extinction in the early 1900s seemed unthinkable. I have written about passenger pigeons before noting that "in the 1800s and before, migratory flocks of passenger pigeons were so immense that they blanketed the skies of eastern North America. One individual recounted a mile wide swath passing overhead blocking the sun for three consecutive days. The birds flew at an estimated speed of about sixty miles an hour. Nesting birds took over entire forests, trees were crammed with dozens of nests, collectively weighing so much that branches would break and tree trunks would topple. Surface vegetation was destroyed by the thick layers of the bird's excrement. Sound was overwhelming. But the birds were not just noisy, they were tasty too, and their arrival guaranteed an abundance of free food."

The greatest current losses have occurred to birds living along shorelines and in grasslands. They have declined by 53 percent since 1970, a loss of 700 million adults in the 31 species studied, including meadowlarks, and northern bobwhites. Shorebirds such as sanderlings and plovers are down by about one-third. Habitat loss may be partly to blame. Of the nearly 3 billion birds lost, 90 percent belong to 12 bird families, including sparrows, warblers, finches and swallows that are common, widespread species that play influential roles in food webs and ecosystem functioning from seed dispersal to pest control.

Evidence from the declines emerged from detection of migratory birds in the air from weather radar stations across the continent in a period spanning over 10 years, as well as from nearly 50 years of data collected through multiple monitoring efforts on the ground.

Although the study did not analyze the causes of the declines, it noted that the steep drop in North American birds parallels the losses of birds elsewhere in the world, suggesting multiple interacting causes that reduce breeding success and increased mortality. The largest factor driving these declines is likely the widespread loss and degradation of habitat, especially due to agriculture and urbanization. Other studies have demonstrated mortality from predators, especially free roaming domestic cats; collisions with glass, buildings, and other structures; and pervasive use of pesticides such as neonicotinoids, associated with widespread declines in insects, an essential food source for birds. Climate change is expected to compound these challenges by altering habitats and threatening plant communities that birds need to survive. More research is needed to pinpoint primary causes for declines in individual species. Until then, the shocking news is a wake up call. According to a recent editorial: "birds are indicator species, serving as acutely sensitive barometers on environmental health, and their mass declines signal that the earth's biological systems are in trouble."

Fortunately, the news is not all bad. There have been a number of promising rebounds resulting from human efforts. Waterfowl (ducks, geese and swans) have made remarkable recovery over the past 50 years, made possible by investments in conservation by hunters and billions of dollars of government funding for wetland protection and restoration. Bald eagles have also made spectacular comebacks since the 1970s, after the harmful DDT was banned (thanks in part to Rachel Carson), and recovery efforts through endangered species legislation in the United States and Canada provided critical protection.

Final thoughts

Each of us can make a difference with every day actions that together can save the lives of millions of birds—actions like making windows safer for birds by adding or keeping screens up year around, keeping cats indoors, planting more trees, keeping fresh water available and protecting habitat. I just donated to the American Bird Conservancy, a nonprofit organization dedicated to conserving birds and their habitats throughout the Americas. There are other organizations as well, including the Bird Conservancy of the Rockies, the Cornell Lab of Ornithology, Advancing Georgetown and the Smithsonian Migratory Bird Center.

Reptiles, Bugs and Marine Life Used for Drugs

Venomous reptiles, bugs and marine life have notorious reputations as dangerous or life threatening creatures, but all can be sources of new classes of drugs capable of treating diabetes, autoimmune diseases, chronic pain and other conditions. To date there have been seven FDA approved venom derived drugs as a result of modern day research. One of the drugs, exenatide, is made from the saliva of the Gila monster and is used in treating type 2 diabetes. Ziconotide is the synthetic equivalent of a peptide found in the venom of a marine snail and used for chronic pain. Captopril is derived from the Brazilian pit viper and used to manage high blood pressure. Eptifibatide and Tirofiban are anti-platelet drugs derived from rattlesnake or viper venoms used to prevent blood clots. Lepirudin and Bivalirudin are thrombin inhibitors synthetic derivatives of hirudin a compound excreted from the medicinal leech.

Purists may argue that some type of venom has been used to treat patients long before either of these products. Snake venom, for example, was used since the 7th century B.C. to prolong life and treat arthritis and gastrointestinal ailments, while tarantulas are and were used in the traditional medicine of indigenous populations of Mexico and Central and South America. There have also been a number of antivenin preparations available for treating snakebite and equine derived antivenin has been the mainstay of hospital treatment for venomous snakebite for more than 35 years. Antivenin is an antibody preparation derived from horses or sheep injected with non-lethal amounts of snake venom. For rattlesnake, cottonmouth, and copperhead bites, for example, Antivenin (*Crotalidae*) Polyvalent has been the standard available treatment. Eastern coral snakebites require Antivenin (*Micrurus fulvius*). A sheep-derived antivenin, CroFab (Crotalidae Polyvalent Immune Fab), has also received approval from the Food and Drug Administration for treating snakebites; it is much less allergenic. Due to their method of manufacture and their passive use, none of the antivenin products can be considered therapeutic venoms. Captopril, the first of a series of ACE inhibitors, however, can be considered a precursor and is the first example of a venom based drug. Captopril was developed from structural relationship studies of teprotide, a peptide from *Bothrops jararaca* venom. Historians may go further back and consider theriac, a remedy developed by Galen in the second century A.D. He developed his concoction containing more than 70 ingredients, including opium, wine and snake meat to protect against snake bite.

The release of exenatide and ziconotide will likely be the impetus for a host of other therapeutically useful drugs. Venoms are proving to be a remarkable source of novel peptides that have potential applications in human health. A number have already been used for proof of concept studies, some having undergone preclinical or clinical development for treating pain, diabetes, multiple sclerosis and cardiovascular diseases. Toxins that target ion channels and receptors have been isolated from spiders, marine snails, snakes, scorpions, and a range of other animals. Snake venoms have been studied extensively, including their molecular origins, chemical nature, and biologic activities. The studies have been aided by new innovations in technologies that map the relationships and actions of the molecular structure of the venom. Researchers now have the ability to uncover evolutionary changes and diversification among specific venomous species that could prove useful in developing new drugs. According to one clinician, knowing more about the evolutionary history of venomous species can help make more targeted decisions about the potential uses of venom compounds in treating disease.

Venoms are made from the same basic molecules as the proteins of the body. They contain short chains of amino acids, called polypeptides, which poison the victim by causing paralysis. Like curare, they act by blocking the action of acetylcholine at the junction between the nerve and muscle. Venomous snakes are known to possess one of the most sophisticated integrated weapons systems in the natural world. Like snakes, Conus venoms also contain a remarkable diversity of pharmacologically active small peptides. Their targets are ion channels and receptors in the neuromuscular system and a drug derived from the venom could treat autoimmune diseases. One noteworthy potential drug is chlorotoxin derived from the deathstalker scorpion. It could be the basis for a surgical tumor-imaging technique and clinical trials to confirm safety and efficacy have already begun. Chlorotoxin, actually lights up or stains (paints) malignant tumors and other cancers and could help surgeons resect tumors with the least amount of extraneous damage to surrounding non-cancerous tissues. With all of the ongoing research there are bound to be a number of surprising new medical developments in store for all of us.

Viruses

—Miraculous life forms.

Viruses may be the most bizarre of all life forms, although they are not truly living. They are however a form of life and consist of mere short pieces of infectious deoxyribonucleic acid (DNA) or ribonucleic acid (RNA) wrapped in a simple protective coat. The famous immunologist, Sir Peter Medawar termed the virus "as a piece of bad news wrapped up in protein." Of course, there is more than just the structure (a virion) to consider, there is an outer coat called the capsid. Capsids come in various sizes and shapes, each characteristic of the virus family to which it belongs. They are built up of protein subunits called capsomeres and it is the arrangement of these around the central genetic material that determines the shape of the virion. Most viruses are too small to be seen under a light microscope as they are about 100 to 500 times smaller than bacteria, varying in size from 20 to 300 nanometers in diameter (one nanometer is a thousand millionth of a meter). Inside the virus capsid is its genetic material, or genome, which is either DNA or RNA depending on the type of virus. The genome contains the virus's genes, which carry the code for making new viruses, and transmits these inherited characteristics to the next generation. Viruses usually have between 4 and 200 genes.

According to Frank Ryan MD, in his book *Virus X,* " viruses have no skin, nerves, and for that matter, no brain, but they do have a way of detecting the chemical composition of targeted cell surfaces." Every virus has a chosen host cell, whether it is the leaf of a tobacco plant in mosaic disease or the T cell in a human sufferer of AIDS. The virus has the most unique ability to sense the right cell surfaces and recognizes them through a perception of three dimensional surface chemistry. They then hijack the targeted cell's components and use what they need, often killing the cell in the process. Thus viruses are obliged to obtain the essential parts of other living things to complete their life cycle. Inside the cell the virus reproduces. The cell eventually bursts open, releasing the virus and the vicious cycle begins anew.

Recent research indicates that many of the viruses infecting us today have ancient evolutionary histories that date back to the first vertebrates and perhaps the first animals in existence. Researchers discovered 214 novel RNA viruses (where genomic material is RNA rather than DNA) in apparently healthy reptiles, amphibians, lungfish, ray-finned and other types of fish. For the first time there is proof that RNA viruses are many million years old, and have been in existence since the first vertebrates existed. The study emphasized how large the universe of viruses really is. Viruses are everywhere and many millions are still to be discovered. It has been estimated that trillions of viruses fall from the sky each day. According to a recent study some 800 million viruses cascade onto every square meter of the planet swept into the air by sea spray, and dust storms. It is assumed that these viruses originate on the planet and swept upward, but it is theorized that viruses may even have come from outer space. Viruses are more than infectious agents, they are essential to everything from our immune system to our gut microbiome, to the ecosystems on land and sea, to climate regulation and the evolution of all species. These surprising life forms contain a vast diverse array of unknown genes and spread them to other species; they are far from simple.

Viruses are associated with plagues—epidemics accompanied by great mortality, such as AIDS, flu and smallpox. The latter may be the most fearsome virus. It was believed to have originated in India in ancient times before first ravaging the Roman world as early as A.D. 165, since then it had scourged humanity in what amounted to a permanent pandemic, causing incalculable loss of life and misery through its morbidity and disfigurement. In its more virulent form caused by the virus *Variola major*, it still caused up to a 50 percent mortality in its victims. As late as 1958, when Russian doctors pressed for a concerted world campaign against it through the World Health Organization, 2 million people still died from its effects each year. A global campaign against smallpox began in 1967 which involved vaccinating as many as 250 million people yearly. It took ten years to achieve success. Physicians at that time felt it possible to eradicate all viral plagues. Today such optimism appears unwarranted as new viruses such as Chikungunya, Lassa, Nipah, or Zika seem to appear almost daily.

VISION

—*What a wonderful gift!*

MANY WILL AGREE that the ability to see is the most critical of the five senses. Few of us, however, may be able to distinguish between sight and vision. Sight is what takes place in your eyes when you see light; vision is what occurs when the messages triggered by that light race through the optic nerve into the visual cortex in the depths of the brain. Eyes are even known to widen in fear, boosting sensitivity and expanding our field of vision to locate surrounding danger. Construction of the human eye and the supporting structures are most remarkable, bewilderingly complicated, elaborately constructed and composed of multiple subsystems.

The entire eyeball is protected by being almost buried in a boney socket. Even the small, exposed front section of the eye is guarded by the eyebrows, eyelashes and the eyelids and the tear apparatus (located under the upper eyelid). The eyebrows help to protect the eyes from dust and foreign bodies. The eyelid keeps out foreign material and then in blinking and closing, spreads the tears across the eye while simultaneously wiping the surface of the eye clean, like the sweep of a windshield wiper. During sleep, the closed eyelid prevents evaporation of moisture in the eye. Eyes cannot survive unless their exposed surfaces are kept continuously moist.

The eye has been defined as an optical device that receives and recognizes light and has the ability to define spatial detail. It is made up a conjunctiva, the lens, the iris and its center aperture the pupil, the cornea, ciliary muscles, sclera, extraocular muscles and the optic nerve. The collective function of the nonretinal parts of the eye is to keep a focused, clear image of the outside world anchored on the retina. The cornea and lens focus light rays onto the back of the eye while the lens regulates the focusing for near and far objects by becoming more or less globular. This change of shape occurs in all mammals, reptiles and birds, in a process of deformation.

Each part of the eye plays a role in how we see. Light bouncing off an object will first contact the cornea. The cornea refracts the rays of light which then passes through the pupil and enters the lens. The lens further bends the light to an image focused on the retina. Light acts like an electric shock through the layers of cells and stimulates them to send their message to the brain. Most of the light is focused on an area of the retina called macula, where vision is the sharpest and clearest.

The normal, healthy human eye has as many as 150 million photoreceptor cells with three visual pigments in the cones for day color vision and one pigment in the rods for night time. The primate retina is complex with many interconnecting neurons, but not as complex as the retina of many birds and even some turtles. While we do not have the best optical device in the animal world, we do have a human brain that helps us meet any deficiencies in retinal construction.

There are a number of afflictions that affect the eye but cataracts are the most common, they affect almost 22 million persons above the age of 40. In its early stages, however, a cataract is not a disease, but a normal part of aging. A cataract is a loss of transparency of the normally clear lens of the eye. Aging causes chemical changes to occur that render it less transparent. The lens opacity causes light to be scattered. As a result, light from an object that should pass directly through the lens produces a degraded image or no image at all.

Unlike other animals, the human lens is yellow, and grows for a remarkably long period of time, increasing from about 90 milligrams at birth to 240 milligrams at age 80. Oxidation damage to the lens is one contributor to cataract formation. The oxidation process releases chemicals called free radicals. In the eye the sources of oxidation include ultraviolet radiation, xray-radiation, and possibly cosmic radiation as well. Because the front of the eye is so transparent, the lens is constantly bathed in light and over many years the delicate protein arrangement in the lens becomes damaged. Other causes include diabetes, where poor control of blood sugar as well as a long duration of the illness play a major role in cataract development. Cataracts are also fairly common in individuals who must take corticosteroids for prolonged periods of time and in cigarette smokers.

Glaucoma is another common eye disease that causes damage to the optic nerve. You should visit your eye care professional, particularly if you are over the 40 and have a family history of the disease.

The eye's evolution from a primitive photosensitive pigment to today's marvelous instrument is indeed a miracle. How fortunate we are to have inherited such a wondrous gift.

Visualizing data

Visual displays of data are evident just about everywhere, they invite people to think about the information in meaningful ways to facilitate understanding and are commonly used in school and in the work place. Many of us have been exposed to PowerPoint or some other graphic method, but few of us have had training in data visualization and bad graphical methods are the result. There are a number of ways to present such information. Graphs are an excellent example of one method for analyzing scientific data and for communicating quantitative information and are commonly used in high school and in college. Most companies use graphs to display financial and sales data.

History

Quantitative graphs, central to the development of science and statistical artwork, date from the earliest attempts to analyze data. The earliest known map found on a clay tablet dated 3800 B.C., depicts all of Northern Mesopotamia with conventions and symbols still familiar today. From about 3200 B.C.. Egyptian surveyors abstracted their lands in terms of coordinates not unlike the Cartesian systems still in use. Statistical graphics including simple tables and plots, date from the earliest attempts to analyze empirical data. Many of the most familiar forms and techniques were well established at least 200 years ago. Today quantitative graphics are re-emerging as an important statistical tool, as evidenced by developments as diverse as the growing use by statisticians of computer graphics, and the proliferation of descriptive graphics in statistical publications of the Federal government.

An earlier issue of the Economist highlighted three of history's best. They included Florence Nightingale's Coxcomb, it came from her monograph on *"Notes on matters affecting the health efficiency and hospital administration of the British Army"*, published in 1858. This was the same year she became the first female fellow of the Statistical Society of London (now Royal Statistical Society). Her map displayed the causes of soldier's deaths during the Crimean war. The second was Charles Joseph Minard's chart that told the story of Napoleon's Russian Campaign of 1812. It displayed six types of information: geography, time, temperature, the course and direction of the army's movement and number of troops remaining, and was published in 1862. His chart has been singled out for special mention: it inspires bitter reflections on the cost to humanity of the madness of conquerors and the merciless thirst for military glory. Anyone with an interest in graphics and history should read Sandra Rendgen's book, *The Minard System*.

The third of history's best came much earlier, it was published in 1821 by William Playfair. One of his charts compared tax levels in various countries to show that Britain's were too high. His most famous displayed the weekly wages of a good mechanic and the price of wheat with the reigns of monarchs featured along the top. Playfair was the first person to use horizontal and vertical axes to present time and money and is the key figure in the history of quantitative graphics. Minard was inspired by Playfair's work.

VISUAL AIDS

If you are given the opportunity to present data in school or in a meeting, certain methods for displaying and analyzing information are better than others. According to Edward Tufte, superior methods are more likely to produce truthful, credible and precise findings. (Tufte is the most famous theorist of information presentation.) The difference between an excellent analysis and a faulty one can sometimes have momentous consequences.

Visual aids help to keep viewers engaged. To make visuals more compelling, keep them simple.

- Minimize the number and complexity of slides.
- Use only three to six ideas on each flip chart or PowerPoint or Prezi slide.
- Use keywords or phrases, not full sentences and use graphics, photos are symbols to reinforce concepts.
- Only one concept and not more than six lines of text per slide should be the standard.
- Use color on slides, but not excessively.

- Use bullet points, not numbers, for nonsequential items.
- Use all upper-case letters only for titles and acronyms.

SOURCES OF EFFECTIVE GRAPHS

Fortunately, there are a number of excellent books available for help in presenting data. One of the best is "Show Me the Numbers", written by Steven Few. His book distinguishes between charts and graphs, describes the use and design of tables, explains visual perception, discusses how to display many variables at once, and how to use graphs readily available in software. Naomi Robbins book on "Creating More Effective Graphs" is easy to read and comprehensive. She provides ways to display more than two variables, statistical software, scales, and a useful checklist of possible graph defects. Perhaps the best books of all were written by Edward Tufte. He is a professor at Yale University, where he teaches courses in statistical evidence and information design. They are beautifully presented and highly recommended and widely acclaimed by a variety of professionals.

Watching worthwhile television

~~~~~

I try not to spend too much time in front of the TV but every once in a while I catch a program worth viewing. One of the shows I am referring to appeared for the first time on Netflix in August, with the title "The Most Unknown." It was a documentary film in the Simons Foundation Science Sandbox series which takes the viewer on a journey with nine scientists as they review and try to understand the universe. The program should appeal to anyone interested in science.

Each of the scientists are from different disciplines and they immerse themselves into another's work and try to solve the most difficult problems. There is proof that such involvement can result in scientific breakthroughs. An earlier study published in the journal Science had indicated that research within narrow boundaries is unlikely to be the source of most fruitful ideas.

The program begins in central Italy, on a deep-cave journey with the geomicrobiologist Jennifer Macalady— "This is probably the most beautiful slime I've ever seen," she says—she then travels to Milan to talk to the particle physicist Davide D'Angelo about dark matter and dark energy. Dark energy affects the expansion of the universe and roughly 68 percent of the universe is dark energy. Dark matter makes up about 27 percent. Normal matter makes up the difference. We are much more certain what dark matter is not than we are what it is. It may be made up of exotic types of particles.

D'Angelo then heads to Brussels, to the lab of the cognitive scientist Axel Cleeremans, who has him strap on an EEG cap and make a robotic hand move with his thoughts.

In one of the segments a microbiologist describes how tiny microbes, too small to be seen with the naked eye can eat greenhouse gases and therefore a potential solution to climate change. These tiny organisms were not recognized as a major domain of life until quite recently. Based on their DNA sequences and their vast difference in genetic makeup they are unlike bacteria and thus are classified as archaea—a new domain.

The scientist then travels to Colorado to meet a physicist obsessed with measuring time so precise that it only loses once second in accuracy every million years. As the two scientists converse, they make a potential connection. The secret to the success of the greenhouse-eating microbes may be their particular experience with time. At the bottom of the ocean, strong gravitational waves bend and slow down time. The change is not large, but it may be the clue to understanding the microbes' specific dietary requirements.

The physicist obsessed with time then travels to meet a neuroscientist who studies why humans can't keep time without reference points of activity. In the laboratory subjects in magnetic resonance imaging machines are asked to guess the run time of video clips. They find that the human perception of time is linked to the activity level in the video. When persons in the video are highly active or move quickly, subjects underestimate the length of the video and just the opposite for low activity videos. One implication is that although our life spans are increasing, social media and frenzied news cycles may lead to the perception that life is shorter and less fulfilling.

One of the discussions deals with the search for the derivation of consciousness—-the spark that makes us *us*. There is something buried inside of all of us that makes us aware of ourselves and our world. Without it we would presumably have no basis for curiosity, no realization that there is a world about which to be curious, on impetus to seek insight, whether emotional, artistic, religious or scientific. Consciousness is tthe window through which we understand.

In one of the last segments, a Yale psychologist is interested in conditions under which monkeys attempt to steal treats from other monkeys and if they act differently with third party witnesses.

The director's goal isn't so much to inform as to inspire, and it's exciting to watch his subjects step out of their own research and into that of their peers. At two different points, scientists giddily say that they feel like a kid on Christmas morning. "Humans get smarter the more things they experience," Macalady says. It's a lesson that sticks with us afterward, out of the theatre and into the world.

The Most Unknown is first rate television and I suggest parents insist or encourage their teen age or even younger children to spent a hour and half in front of the television watching and learning. After viewing, you might even wish to spend some time discussing the topics with them. There are few better ways to learn how science explains the universe, the world we live in, and the worlds within us.

WATER

~~~~

WHEN THE VERY first known Greek scientist, philosopher and mathematician, Thales of Miletus (546-624 BC), was asked to explain the world, he unexpectedly answered: water. Water he said can change its form from solid to liquid to gas and back again; clouds and rivers were in essence water, and water was essential for life. He, of course, was correct. Fortunately, water is ubiquitous, it makes up about 70 per cent of the earth's surface and about the same percentage of our body weight. Despite water's obvious value, most of us know little about its physical, chemical, health or medicinal properties. We are not even sure how water formed on earth. It may have arrived here via comets or meteorites or formed by reaction between some of the released gases in the early atmosphere. The subject is fascinating and according to an editorial in the British Medical Journal, water has become a modern fashion and health accessory. Students have bottles of water in their school bags, people are jogging with water, and office workers have bottled water within easy access of their desks. We spend billions on bottled water worldwide. Strangely, we older people survived even without the endless supply of water bottles that nearly everyone seems to carry around these days.

Of course, nothing on earth can replace water. In fact, when astronomers search for life in outer space, they look for water. Without exception, every form of life ever found requires water.

According to the classical Chinese text Tao Te Ching, "The highest excellence is like water. There is nothing in the world softer and weaker than water, and yet, when it comes to attacking things that are firm and strong there is nothing that can surpass it—because there is nothing that is so effective that it can replace water." Water is also one of the few compounds with a higher density as a liquid than a solid, this means that heavier water displaces the lighter ice and ice will float.

In most aspects of our health, we are taught that moderation is the key. But the message when it comes to water has always been more, more, more. In our quest to be healthy, we've always thought we could never drink too much water. The conventional wisdom was half right. Proper hydration is the key to unlocking optimal health. But we need to start looking at hydration for what it is: the very essence of your health. You are a body of water. In fact, by the most modest traditional estimates, approximately 65 percent of you is water. (From 75 percent in infants to 55 percent in the elderly.) If you're not hydrated, everything else you do to stay healthy (exercising, eating right, stress management, sleep) is undercut. It's known that humans can survive for about two months without food, but just days without water will kill us. Inadequate hydration can cause fatigue, poor appetite, heat intolerance, dizziness, constipation, kidney stones, urinary tract infections and a dangerous drop in blood pressure. Other disorders resulting from dehydration may come as a surprise: sleeplessness, decreased immunity, joint pain, chronic diseases like fibromyalgia.

According to Dana Cohen and Gina Bria's recent book *Quench,* every bodily process, every living cell, depends on water to function properly. Water transports nutrients, regulates body temperature, lubricates joints and internal organs, supports the structure of cells and tissues and preserves cardiovascular function.

Jane Brody, New York Times health columnist, reports that the typical American consumes about a liter or a little over four cups of drinking water a day. Anyone engaged in vigorous physical activity should drink more. If you plan to do strenuous exercise or physical activity outdoors on a hot day, it is better to start hydrating the day before. Check the color of your urine, the paler, the better. Continue to drink water or other fluids throughout your activity and for hours afterward. To remain well hydrated, do not rely on thirst to remind you to drink, be proactive by consuming enough liquid before, during, and after meals and physical activity. The long standing advice to drink eight glasses of water a day has been modified. Current thinking calls for obtaining 70 percent of daily water needs from liquids (including coffee or tea, not alcohol) and the rest from solid foods. The authors of Quench suggest two dozen fruits and vegetables that are especially hydrating, ranging from cucumbers (96.7 percent water) to grapes (81.5 percent). Other fruits and vegetables to consider includes lettuce, tomatoes, cauliflower, spinach, broccoli, carrots, peppers, watermelon, strawberries, pineapple, blueberries, apples and pears. Naturally packaged plant water hydrates more efficiently than plain drinking water because it is already purified, is packed with soluble nutrients and gradually supplies the body with water.

According to the National Academies of Science, the vast majority of healthy people meet their daily hydration needs by letting thirst be their guide. The bottom line is to drink when you're thirsty and more when you sweat. Your body will take it from there.

What is science?

Today, if someone would ask a young person to define "science", chances are he or she would Google the word or ask Siri or some other personal assistant for the answer. (Siri is available on Apple devices capable of responding to queries.) Seemingly, Information comes relatively easy these days, but with drawbacks. The main one is lack of confirmation as to authenticity, the second is gullibility, the third and most important is losing the ability to think and reason. During these times of allegedly readily information it is wise to consider the Latin phrase, *Nullius in Verba*, translated as "on the word of no one" or" take no one's word for it." It is the motto of the Royal Society in the United Kingdom. Checking and verifying the source and accuracy of the information obtained on the internet is extremely important.

The word "science" is derived from the Latin word *scientia*, which is knowledge based on demonstrable and reproducible data, according to the Merriam-Webster Dictionary. True to this definition, science aims for measurable results through testing and analysis. Science is based on fact, not opinion or preferences. The process of science is designed to challenge ideas through research. The Encyclopedia Brittanica defines science as any system of knowledge that is concerned with the physical world and its phenomena and that entails unbiased observations and systematic experimentation. In general, a science involves a pursuit of knowledge covering general truths or the operations of fundamental laws." Perhaps there is no one better than Edward O. Wilson for explaining science. (Dr. Wilson is an American biologist, naturalist and writer and world's leading expert on ants.) In one of his many books, *Letters to a Young Scientist*, he wrote: "It is organized, testable knowledge of the real world, of everything around us as well as ourselves, as opposed to the endlessly varied beliefs people hold from myth and superstition. It is the combination of physical and mental operations that have become increasingly the habit of educated peoples, a culture of illuminations dedicated to the most effective way ever conceived of acquiring factual knowledge." Few scholars could define it as well. The book itself should be compulsory reading for high school students with an interest in science.

According to Richard Giere, in his book *Explaining Science*, "Science is more than just the definition. It is a cognitive activity, which is to say it is concerned with the generation of knowledge. Indeed, science is now the major paradigm of a knowledge-producing enterprise." He describes science as both a body of knowledge and a process. In school, science may sometimes seem like a collection of isolated and static facts listed in a textbook, but that's only a small part of the story. Just as importantly, science is also a process of discovery that allows us to link isolated facts into coherent and comprehensive understandings of the natural world. Science is exciting. Science is a way of discovering what's in the universe and how those things work today, how they worked in the past, and how they are likely to work in the future. Scientists are motivated by the thrill of seeing or figuring out something that no one has before. Science is useful. The knowledge generated by science is powerful and reliable. It can be used to develop new technologies, treat diseases, and deal with many other sorts of problems. Science is ongoing. Science is continually refining and expanding our knowledge of the universe, and as it does, it leads to new questions for future investigation. Science will never be "finished." Science is a global human endeavor. People all over the world participate in the process of science.

Science literacy has pervaded the purposes of school science since at least the 1990s. It reflects concerns that school science should prepare citizens to engage with science as well as prepare for science-related careers. Science literacy refers to a student's capacity to master the literacy practices of science, which enable them to conduct investigations, collect and interpret data, critique claims and make informed decisions.

Final thoughts

In case you as parents have young people at home who complain about taking courses at school with little relevance to their life's work. Tell them how important science is in problem solving and decision making. Many of the major challenges and opportunities that confront us need to be approached from a scientific perspective, as our recent experience with the corona virus tells us. By studying science, students develop an understanding of the world, learn that science involves particular processes as ways of developing and organizing knowledge for future research.

Zebrafish

—An amazing tool for biomedical research

Those humble fish swimming gracefully in your aquarium have become excellent models for studying childhood cancer and a host of other diseases. Over the past decade they have proven to be an important tool for many areas of biomedical research, including drug discoveries, cardiovascular and neurological disease, aging and toxicology. Zebrafish models of infectious diseases such as tuberculosis have been established that are now amenable to high throughput *in vivo* drug screens, a much needed development in the fight against drug-resistant microorganisms. Zebrafish larvae have even been used to help in the search for appetite suppressants. This involves feeding fluorescent organisms to the fish in order to quantify their feeding behavior—the more fluorescence in the larvae's stomach, the larger their appetite. Drugs can be administered to detect possible unwanted side effects on the larvae's behavior. Nicotine, for example, was found to reduce the larvae's appetite. Fluorescent tagging zebrafish has been employed as a powerful way to find new cancer drugs. Just recently scientists have identified brain cells vital to how zebrafish socialize. When their neurons are disabled, orientation to one another breaks down in ways similar to socialization problems seen in humans with autism and schizophrenia.

The **zebrafish**, ***Danio rerio***, is a tropical freshwater fish belonging to the minnow family (Cyprinidae) of order Cypriniformes. It is a popular aquarium fish, frequently sold under the trade name **zebra danio.** Zebrafish were originally found in slow streams and rice paddies and in the Ganges River in East India and Burma.

In the early 1970's, Dr. George Streisinger (1927-1984), a scientist at the University of Oregon, determined that the zebrafish was an ideal model for studying vertebrate development and genetics. Dr. Streisinger is considered by many of his peers as the founding father of zebrafish research. He recognized that the short generation time of the zebrafish (2-3 months), its high birth rate (100-200 embryos per mating) and oviparous (producing eggs in which the embryo develops outside of the maternal body) mode of reproduction were ideally suited for the rapid screening of the progeny of mutagenized fish for mutations that affect important developmental processes. His proven use of the zebrafish in research has spread to over 300 developmental and genetic laboratories in over 30 countries and many of the mutant strains produced in his lab are still alive in research facilities throughout the world and used to provide answers to human and animal health issues.

The zebrafish is the first vertebrate proven tractable to large scale genetic screening formerly used so successfully in fruit flies and nematodes (worms). It has many features that make it an excellent model organism for studying development in vertebrates. The embryos develop externally to the mother and are transparent, so they can be easily viewed and manipulated. The transparency of the embryos has become one of the leading reasons for using them. Being able to see through the embryos allows researchers to watch the morphological changes that occur during development.

As mentioned above, the zebrafish has several advantages as a model for studying vertebrate developmental processes. These include small size, easy care, and rapid generation time. In addition, the embryo develops from eggs that are externally fertilized. Because of their transparency the embryos can be continuously observed under light microscopy. Mutagenesis screens can detect defects in early organogenesis and late organ function. Because the fish is a vertebrate its genetic program is more similar to that of mammals than invertebrate models. The evolutionary divergence of fish from the mammalian lineage occurred roughly 300 million years ago.

New studies have indicated that zebrafish can be used to discover psychoactive drugs. In the past in vitro screening assays could not be used to recreate the complex network interactions of whole organisms and thus it was not possible to predict how small molecules would alter complex behaviors. Recently discovered methods can now employ zebrafish developmental processes as a high throughput screen for small molecules that are designed to alter larval zebrafish locomotor behavior.

Recent experiments have also discovered Zebrafish's remarkable ability to mend a damaged heart. (Fortunately, the hearts of zebrafish and humans both have chambers and rhythmically pump oxygen carrying blood through the body.) Mature cardiocytes (cardiac muscle cells) near the injury site detach from one another and lose their typical shape to make it possible to start dividing again to replace lost tissue. Within two weeks, the new heart tissue can receive electrical signals in the same manner as normal healthy heart tissue. Figuring out exactly how zebrafish accomplish their self repair could help scientists find ways to trigger similar regeneration in human patients.

We can thank this tiny fish for making such a big difference to our health.

www.ingramcontent.com/pod-product-compliance
Lightning Source LLC
Chambersburg PA
CBHW031603210526
45464CB00004B/1403